Cosmic Purpose

VERITAS

Series Introduction

"... the truth will set you free." (John 8:32)

In much contemporary discourse, Pilate's question has been taken to mark the absolute boundary of human thought. Beyond this boundary, it is often suggested, is an intellectual hinterland into which we must not venture. This terrain is an agnosticism of thought: because truth cannot be possessed, it must not be spoken. Thus, it is argued that the defenders of "truth" in our day are often traffickers in ideology, merchants of counterfeits, or anti-liberal. They are, because it is somewhat taken for granted that Nietzsche's word is final: truth is the domain of tyranny.

Is this indeed the case, or might another vision of truth offer itself? The ancient Greeks named the love of wisdom as *philia*, or friendship. The one who would become wise, they argued, would be a "friend of truth." For both philosophy and theology might be conceived as schools in the friendship of truth, as a kind of relation. For like friendship, truth is as much discovered as it is made. If truth is then so elusive, if its domain is *terra incognita*, perhaps this is because it arrives to us—unannounced—as gift, as a person, and not some thing.

The aim of the Veritas book series is to publish incisive and original current scholarly work that inhabits "the between" and "the beyond" of theology and philosophy. These volumes will all share a common aspiration to transcend the institutional divorce in which these two disciplines often find themselves, and to engage questions of pressing concern to both philosophers and theologians in such a way as to reinvigorate both disciples with a kind of interdisciplinary desire, often so absent in contemporary academe. In a word, these volumes represent collective efforts in the befriending of truth, doing so beyond the simulacra of pretend tolerance, the violent, yet insipid reasoning of liberalism that asks with Pilate, "What is truth?"—expecting a consensus of non-commitment; one that encourages the commodification of the mind, now sedated by the civil service of career, ministered by the frightened patrons of position.

The series will therefore consist of two "wings": (1) original monographs; and (2) essay collections on a range of topics in theology and philosophy. The latter will principally by the products of the annual conferences of the Centre of Theology and Philosophy (www.theologyphilosophycentre.co.uk).

Conor Cunningham
Eric Austin Lee
Series editors

Cosmic Purpose

Kagawa Toyohiko

edited by
Thomas John Hastings

translated by
James W. Heisig

CASCADE *Books* • Eugene, Oregon

COSMIC PURPOSE

Veritas 12

Copyright © 2014 Thomas John Hastings. All rights reserved. Except for brief quotations in critical publications or reviews, no part of this book may be reproduced in any manner without prior written permission from the publisher. Write: Permissions, Wipf and Stock Publishers, 199 W. 8th Ave., Suite 3, Eugene, OR 97401.

> This publication was made possible through the support of a grant from the John Templeton Foundation. The opinions expressed in this publication are those of the author and editor and do not necessarily reflect the views of the John Templeton Foundation.

Cascade Books
A Division of Wipf and Stock Publishers
199 W. 8th Ave., Suite 3
Eugene, OR 97401

www.wipfandstock.com

ISBN 13: 978-1-62564-509-8

Cataloging-in-Publication data:

Cosmic Purpose / edited by Thomas John Hastings and translated by James W. Heisig.

Veritas 12

vi + 275 p. ; 23 cm. Includes index.

ISBN 13: 978-1-62564-509-8

1. Religion and science. 2. Evolution (Biology). 3. Evolution—Religious aspects—Christianity.

I. Hastings, Thomas John. II. Heisig, James W., 1944–. III. Series. IV. Title.

T712 K25 2014

Manufactured in the U.S.A.

Contents

Editor's Introduction   1
Acknowledgements   27

Preface   29

DIRECTIONALITY IN NATURAL SELECTION

1. Natural Selection and Directionality   31
    The principle of natural selection   41
    Opening and closing in natural selection   48
    Laws governing selection   50
    Oparin's materialism   27
    Statistical mysteries   28
    The development of a teleology of selection   31

2. Standardization in Natural Selection   66
    *Integral standards in natural phenomena and the principle
        of natural selection*   67
    Integral ratios and design   77
    *Yukawa Hideki's theory of the elements and standards
        in the universe*   90
    The direction of a priori selection   93

3. The Fitness of the Natural Environment   104
    Henderson's theory of selective tendencies   104
    Tendency in water   109
    Finality in rocks   117
    The fitness of soil for life-forms   122

LATENT PURPOSE IN THE STRUCTURE OF LIFE

4. Latent Purpose in the Structure of Organisms   131
    Adaptation in the inner environment of organisms   131
    The finality of enzymes   134
    Generating electricity through proteins   138
    Adaptation in physiologically free mechanisms   144
    The phenomenon of life as seen in muscle activity   148
    The appearance of organizers in living bodies   150
    Adaptability in the mechanism for nourishment   153

v

5. Adaptation in the Inner Environment   157
    Adaptation in the inner environment of living things   157
    Dürkhen's teleology of life   162
    Adaptation in the genetic mechanisms   166

6. Attuning to the Struggle for Survival   178
    Adjustment in the struggle for survival   178

THE ESSENCE OF COSMIC PURPOSE

7. Knowledge of Cosmic Purpose   189
    The structure of purpose   192
    A resolution of Kant's antinomies   198
    Chance and purpose: the rise of cosmic evil   205
    Mechanism and purpose   214
    Absoluteness in purpose   227
    Types of purpose   228
    Directionality and purposive design   229
    The many faces of purpose   232

8. The Emergence of Self-Conscious Purpose   246
    The unfolding of self-conscious purpose   246
    Darwin's teleology   252
    The unfolding of conscious purpose   255

9. Cosmic Evil and Its Salvation   263
    The final end of creative cosmic evolution   263
    The dawn of the universe   265
    Cosmic evil and its salvation   267

Index of Personal Names   271

Editor's Introduction

As the "life work" of the once renowned Japanese pastor, evangelist, social reformer, and writer Kagawa Toyohiko (1888–1960),[1] *Cosmic Purpose* does not fit easily into any recognizable contemporary genre. In part one of his three-part commentary on this unique book, the Roman Catholic scholar Kishi Hideshi refers to Kagawa as "the sole cosmological thinker in Japan," who has constructed in *Cosmic Purpose* an "evolutionary doctrine of creation."[2] On a first reading, any number of questions are likely to arise in the minds of readers. With whom is the author quarreling and why? What interpretive framework is he employing in this sweeping argument for finality? A brief account of the historical background, motivations, and conceptual framework behind the book may help answer such questions and provide a general guide to Kagawa's *magnum opus*. From there we shall comment on the initial reception of Kagawa's book

---

1. The *Collected Works of Kagawa Toyohiko*『賀川豊彦全集』were originally published in 24 volumes from 1962 to 1964 and reissued twice, from 1973 to 1974 and from 1981 to 1983. Thematically, we may outline the *Collected Works* as following:
   1. God, Christ, Scripture, and Education: 43 books in 7 volumes.
   2. Economics, Philosophy, Science, and Sociology: 36 books in 6 volumes.
   3. Literature I (novels): 19 books in 6 volumes.
   4. Literature II (poetry and children's stories): 13 books in 1 volume.
   5. Literature III (essays, travelogues, miscellaneous writings in 4 volumes).

Including the books he penned himself, Kagawa actually published some 316 books during his lifetime. This was due to the help of his so-called "Five Pens," dedicated collaborators who took notes either from direct dictation or lectures, compiling and presenting them to Kagawa for final approval. In a real sense, he and his colleagues operated a cottage writing industry, using the proceeds from publications to fund the many projects he and his wife and co-worker Haru and their supporters administered. Only twenty of Kagawa's written works—mainly novels and spiritual writings—were translated into English editions that have since gone out of print.

2. Kishi Hideshi 岸英司,「『宇宙の目的』理解のために (1) [Understanding *Cosmic Purpose* (1)] and (3)『賀川豊彦研究』[Kagawa Toyohiko research] (Toyko: Honjo Kagawa Kinenkan, 1987), 2, 8.

1

in Japan, critically address a number of issues, and consider its relevance for us today.

SCIENCE, RELIGION, MODERNITY

Kagawa came of age during the period of accelerated industrialization and militarization that supported Japan's rise as East Asia's first modern nation state during the late Meiji (1890–1912) and Taishō (1912–1926) Eras. This period of nation and empire building was also a time of ideological confusion and innovation, witnessing the dissolution of many traditional institutions and the advent of increasingly specialized ones. Not surprisingly, Japan's leap into modernity generated a host of new problems.

Keenly sensitized to the psychological, social, and cultural price Japan had to pay for this transition as well as to the opportunities it presented, Kagawa dedicated himself tirelessly to the spiritual and material renewal of Japanese society. Already from his youth he displayed a broad and voracious appetite for reading across the sciences and humanities. Although he opted for the life of a Christian minister and social reformer over that of a teacher and academic researcher, he was well known as a public intellectual, speaker, and essayist engaged in debates on contemporary issues. A year after the end of World War I, he published a book entitled *Spiritual Movements and Social Movements* that reveals his sensitivity to the presence of what he calls "cosmic evil" in society. In the Preface, he writes:

> I am certainly not satisfied with the collection of essays published here. In the near future, I hope to gather my thoughts on cosmic evil into a more systematic form. Yet, I am afraid such a synthesis may take a long time.... The eight million souls sacrificed in Europe cry out in anguish for the inauguration of a new age. I must organize my own reflections for the sake of this coming age![3]

Japan had avoided major mobilization on behalf of the Entente powers to which it was allied, but Kagawa increasingly found himself embroiled in bitter conflicts both with labor leaders advocating imminent revolution and with conservative church leaders encouraging an individualistic piety. In response, he proposed a middle way that would combine social

---

3. 「精神運動と社会運動」 [Spiritual movements and social movements] *Collected Works of Kagawa Toyohiko*, 8/24: 261.

and spiritual movements as public partners in the creation of a robust modern society.

For Kishi, Kagawa's broad middle way represented a "theory of cosmic beauty," that "stresses the evolution of beauty penetrating plants, animals, and human beings."[4] Kagawa's aesthetic insight arose out of serious reflection on Darwin's theory of evolution, Bergson's vitalism, and religious and philosophical approaches to the "cosmic will." As he puts it:

> The aesthetic impulse at work in the cosmic will ascends all the way to human beings.... Beauty should be sought and grasped as the impulsive reality underlying the cosmic will."[5]

In line with his gradualist and holistic answer to the threats that accompanied Japan's headlong plunge into modernization, Kagawa pursued a broad synthesis aimed at holding together the material and spiritual, contingent and unconditional, random and intentional, continuous and discontinuous, and empirical and non-empirical dimensions of the biological, psychological, social, and cultural facets of life. Given the wide-ranging approach of what we may call his "religio-aesthetic cosmic synthesis," it is hardly surprising to find Kagawa calling for a sustained collaboration between the natural sciences and human sciences at the very time that academic disciplines were becoming more and more specialized and cut off from each other. As he notes in *Meditations on God* (1930), "Religion should be explored in terms of the humanities, but it also embraces the cosmology of the natural sciences."[6] In an age when increasing specialization was accepted as an inevitable consequence of the modern explosion of knowledge, Kagawa strived to see all things as a relational unity that, in Einstein's terms, coheres in some "deep but not devious way."

Convinced of the rationality of evolution, Kagawa recognized that all knowledge of the human person, the natural world, and ultimate reality was itself evolving. Like Teilhard de Chardin, he held to the ongoing evolution of spirit and therefore also of religion. Kishi compares the two evolutionary Christian thinkers:

> Teilhard and Kagawa were men not only of the same era but of the

---

4. Kishi, "Understanding *Cosmic Purpose* (1)," 3.
5. Kagawa, *Spiritual Movements and Social Movements*, 325–6, quoted in Kishi.
6. 「神に就いての瞑想」 [Meditations on God], *Collected Works of Kagawa Toyohiko*, 3/24: 12.

Editor's Introduction

same mind. Both saw the cosmos as a stage on which the drama of a mysterious (or mystical) evolution is being played. At the culmination of cosmic evolution Teilhard places the "Omega point," Kagawa "purpose."[7]

While Kagawa grappled with the significance of developments in the natural sciences for religion, he also questioned the significance of fresh religious insight for the sciences. In the opening pages of *Meditations on God* he refers to Galileo, Schleiermacher, Hegel, Comte, Marx, Strauss, Haeckel, Lombroso, Eucken, Kropotkin, Bergson, and Russell, concluding that "whereas faith teaches us about the fate of humanity, science teaches us about the structure of the universe. Those who fail to recognize this distinction can only end up in a great contradiction."[8] He accepted the modernist division of labor for religion and science but not without insisting on the need for interdisciplinary work. "We must pay closer attention to the complexities of the world and preserve the habit of gazing at the universe from different axes."[9]

Believing that the sciences and religions of humanity would have much to gain from taking each other more seriously, Kagawa was one of those rare thinkers who dared to "see things whole" in an age when things seemed to be falling apart. On his first return to the United States after World War II, he offered an illuminating look at his driving spirit:

> I am a scientific mystic. The more scientific I am, the more I feel that I am penetrating deeply into God's world. Especially in the domain of biology do I feel as though I am talking face to face with God. Through life, I discover purpose even in a mechanical world. Science is the mystery of mysteries. It is the divine revelation of revelations.[10]

A positive fusion of his native spiritual and philosophical heritage, his adopted Christian faith, and his interest in modern science transformed Kagawa's innate capacity for "seeing things whole" into a habitual disposition that affected everything he wrote. He acknowledges this tripartite

---

7. Hideshi Kishi, "The Religious Aspects of Cosmic Consciousness: A Comparison of Pierre Teilhard de Chardin and Toyohiko Kagawa," *Christian Century*, December 23, 1970, 1536.

8. Kagawa, *Meditations on God*, 5.

9. Ibid., 22.

10. Emerson O. Bradshaw, *Unconquerable Kagawa* (St. Paul: Macalester Park, 1952), 57–8.

synthesis of Asian emptiness, European theism, and modern science at the conclusion of *Cosmic Purpose*:

> From ancient times people have set out to explain salvation from cosmic evil in one of three ways. First is India's religious way, the idea of emptiness. Second is the theistic approach to salvation that developed in Western European thought. Third is the modern scientific attempt to banish cosmic evil.
>
> I do not find these three to be incompatible. Each of them was bred in human consciousness. Nishida Kitarō recognized the conscious efficacy of the idea of "nothingness." In the Middle Ages, Nicholas of Cusa acknowledged "zero" algebraically. The modern quantum mechanic physicist Hermann Weyl has followed the same line of thought. We are right to eliminate the idea of a meaningless void, but I am speaking of opting for "zero" as a way to think of removing cosmic evil. Moreover, the third path of science's banishment of evil, in its modern meaning, also requires our utmost efforts.
>
> There are, however, limits to human strength that leave us no other solution than to recognize the dependence of everything on an absolute cosmic will that has prepared, a priori, the strength for human beings to survive and for evolution to develop.[11]

It bears keeping this self-characterization and tripartite synthesis in mind while plowing through the dense technical parts of *Cosmic Purpose*. Despite its imposing range of scientific discovery and insight, its enduring value does not lie in its science. Rather, it represents a lifetime of deliberations, some more compelling than others, by one of the most remarkable religious figures of the twentieth century, on the perennial quest for purpose in a vast, beautiful, and mysterious cosmos where shadow, suffering, and evil exists alongside light, flourishing, and goodness. *Cosmic Purpose* is a sustained religious meditation on the implications of contemporary findings in natural science.

A LIFE OF CHRISTIAN PRAXIS

The publication of the book came as something of a shock to Japanese readers. Kagawa was known as a practical-minded Christian evangelist and social reformer. He had lived and worked for ten years in the worst slum of Kōbe, where he came to know firsthand the misery brought about by Japan's rapid industrialization and urbanization. What

---

11. *Cosmic Purpose*, 269.

had made Kagawa a celebrated figure at home and abroad was not his "cosmic synthesis" but his sincere desire to imitate Jesus' love for the poor.

His first public expression of this spiritual longing came in *Across the Death-line*,[12] an autobiographical novel dealing with his traumatic childhood, brooding adolescence, conversion to Christian faith, and life in the slum. To his great surprise, the novel caught the attention of Japanese readers. Within a few years of publication it had sold over a million copies and made him an instant celebrity. Kagawa's circle distributed all of the novel's royalties among their many social projects.[13] The story of his dramatic decision in 1909 to leave his seminary dormitory to live with the poor lent authenticity and authority to his aspirations to a life of awakened love.

Kagawa's theological views of the redemptive love of Christ are creative but hardly *sui generis* or unorthodox. His aim was to synthesize the moral aspects of theories of atonement, *unio mystica*, *theologia cruxis*, and Anglo-American expressions of the social gospel, anticipating in some ways later developments in liberation theology. The cross of Christ was the central symbol for Kagawa, not as a distant historical event or a forensic transaction between sinful individuals and a vengeful God whose honor needed restoring. For him the cross was a "principle of cosmic repair" and a direct expression of the "consciousness of collective responsibility."[14] What Christ embodied in his person does not belong to the past but must continue to inspire those who actively devote themselves to mending what is broken in the individual, society, culture, and the natural world. Kagawa was critical of the "pulpit Christianity" he saw as the dominant form of Protestantism in Japan. He believed the witness of Christianity as a religious minority in Japanese society should have

---

12. Kagawa Toyohiko 賀川豊彦,『死線を越えて』[Across the death-line] (Tokyo: Kaizōsha, 1920). Two English translations were soon released, making Kagawa known abroad as well as in Japan. See Toyohiko Kagawa, *Across the Death-Line* (Kōbe: Japan Chronicle, 1922) and Toyohiko Kagawa, *Before the Dawn* (New York: George H. Doran, 1924).

13. Hamada Yō 濱田陽、「賀川豊彦の遺産と現代」[The legacy of Kagawa Toyohiko and modernity]『日本宗教学会発表資料』[Presentations of the Japanese Association for Religious Studies] September, 2008. Kagawa's wife and co-worker Haru played a vital role in all of these projects. See Mihara Yōko 三原容子,「愛妻—ハルの幸い、社会の幸い」[Beloved wife: The happiness of Haru and the happiness of society]『Think Kagawa—ともに生きる』(Tokyo: Kagawa Archives and Resource Center, 2010).

14.「宇宙創造と人生再創造」[The creation of the universe and the recreation of human life] *Collected Works of Kagawa Toyohiko*, 4/24.

more to do with concrete acts of life-giving love than with the abstractions of theological orthodoxy. His formative years in the slum may be seen as both an expression and a test of his praxis-oriented interpretation of the cross and the *imitatio Christi*. It is little wonder that he was viewed more as a religious practitioner than as a thinker.

As Kagawa was slowly working out the "cosmic synthesis" that would find a final expression in *Cosmic Purpose*, he came to see his own life as an artistic integration of his upbringing in the Shinto, Buddhist, and Neo-Confucian *habitus* of Japan with reflections on Christian faith, growing commitment to social reform, and broad reading in the sciences and humanities. Any attempt to uncover the psychological roots to this synthetic impulse cannot fail to include his experiences as a young boy. At the age of four, he had suffered in quick succession the deaths of his father, Jun'ichi, and Kame, his birth mother and father's concubine. This was followed by an unhappy childhood and a gloomy adolescence with his father's legal wife and other relatives, from which his only solace was nature and books.[15]

The Promethean quest of his life for a "cosmic synthesis" may in part represent an attempt to recover the basic trust and sense of belonging that had been threatened by the traumas of his youth. In language that seems particularly well-suited to Kagawa's development as a *homo religiosus*, James Loder finds deep religious significance in the face of the primary caregiver as the "cosmic ordering, self-confirming presence of a loving other."[16]

THE LOGIC OF FINALITY

Along with this search for a "cosmic synthesis," as early as *The Religion of Life and the Art of Life* (1922)[17] Kagawa began to formulate his core theoretical insight, the so-called "logic of finality" elaborated in *Cosmic Purpose*.[18] This idea, a subspecies of his "cosmic synthesis," grew

---

15. Amemiya Eiichi 雨宮栄一,『青春の賀川豊彦』[Kagawa Toyohiko's youth] (Tokyo: Shinkyō Shuppansha, 2003).

16. James E. Loder, *The Logic of the Spirit: Human Development in Theological Perspective* (San Francisco: Jossey-Bass, 1989), 90.

17. Kagawa Toyohiko賀川豊彦,「生命宗教と生命芸術」[The religion of life and the art of living], *Collected Works of Kagawa Toyohiko*, 4/24

18. *Cosmic Purpose*, 257.

out of his struggles with "cosmic evil." In the Preface to *Cosmic Purpose*, he recalls:

> Shortly before the outbreak of the Pacific War, I began to reconsider the problem of cosmic evil from the perspective of cosmic purpose, which brought me to a new, artistic interest in the structure of the universe. I had a deep sense that the mysterious unfolding of the structure of the cosmos was still in process. Without rushing to any conclusions, I felt the need to focus on the grand production of the universe.[19]

Later in the book he argues that a synthetic approach is preferable to an analytical one if one hopes to grasp the presence of cosmic purpose: "When disassembled and analyzed, purpose becomes incomprehensible. Purpose is synthetic and only becomes clear in its actual assembly."[20] The logic of finality serves as a kind of interpretive key to *Cosmic Purpose*, allowing him to describe the purposive "assembly" he finds at work throughout the universe. In particular, it enables him to connect *life* to *purpose* through an alignment of progressive stages or movements that appear together or separately throughout the book:

life → *energy* → *change* → *growth* → *selection* → *law* → *purpose*

The terms themselves are clearly indebted to Darwin's theory of natural selection, but Kagawa's "logic" is actually an attempt to break through natural selection to a teleological interpretation. In particular, he takes issue with Darwin on the relation between selection and purpose, noting that "Darwin acknowledged selection but denied purpose. I consider this a fundamental contradiction in his thought."[21] Kishi has this to say on the point:

> The main concept in Kagawa's *Cosmic Purpose* is "the directionality of natural selection." For Kagawa, evolution has a direction that moves toward a purpose. The complexity that occurs within selection, that is, the complexity of selection, generates a movement from matter to life and from life to consciousness.[22]

Kagawa seeks to overcome the alleged contradiction in Darwin by offering accumulative evidence for his logic of finality in the intricate

---

19. *Ibid.*, 29.
20. *Ibid.*, 196.
21. *Ibid.*, 215.
22. Kishi, "Understanding Cosmic Purpose (1)," 7.

operations of selection that physics, mathematics, astronomy, astrophysics, chemistry, mineralogy, biochemistry, genetics, biology, physiology, and psychology show to be present at every level of reality.

Although Kagawa alludes to the logic of finality early on, he does not explain it until the last part of the book. It may be that he assumed Japanese readers would recognize the idea from his earlier writings and lectures.[23] Be that as it may, in Part Three, "Knowledge of Cosmic Purpose," we find the clearest statement of how the logic of finality functions as an integrated and purposive structure:

> In order to realize a purpose, simple though it might be, five elements are necessary: energy, change, growth, selection, and law. These five are complementary, but it is energy that first sets the pursuit of purpose on its way. Energy accepts change, grows into purpose, eliminates all sorts of obstacles, selects the best path and method for the purpose, preserves adjustment to the surrounding conditions and environment, and protects the various arrangements and conditions that allow it to ascend to its goal. These five factors together manifest the *power of life* within the living world.[24]

These lines help clarify some of the more esoteric examples that appear in earlier sections of the work, as when he sees the logic of finality at work in chemistry, astrophysics, and biology:

> From the standpoint of ligand field theory, the sun is plus and the earth is minus. Or again, taking the constellation Sagittarius as a galaxy in the solar system, if the surrounding area is a plus and revolves around it, then we should think of it as a moving thing with a minus form. As simple as this sounds, the organization of the cosmic mechanism through simple selection is unimaginably complex. In terms of *energy*, simple selection becomes directionality, the "+" and "–" of axial selection. *Change* involves the plus and minus of chemical combination; *development*, a choice between moving forward and standing still. In the world of *life* sexual differentiation appears; in the world of *law*, a choice between affirmation and negation comes into play. Here again, as simple as this may seem, when we combine the work that goes on at the level of energy, change, development, law, and

---

23. For example, in *Meditations on the Cross* (1931), he refers to "the logic of finality" as "the seven basic elements of truth" and utilizes it to interpret the meaning of Christ's cross.「十字架に就いての瞑想」, *Collected Works of Kagawa Toyohiko*, 3. English translation, *Meditations on the Cross*, (Chicago: Willet, Clark & Company, 1935), 96–7.

24. *Cosmic Purpose*, 192.

life with simple selection, we have a world whose complexity defies imagination.[25]

SCIENTIFIC BACKDROP

In the midst of the storm gathering after the Manchurian Incident of 1931 and Japan's withdrawal from the League of Nations in 1933, Kagawa collaborated with Nakamura Shishio on a translation he had prepared of Sir James Jeans's *The New Background of Science* (1933). In the Preface he expressed his great excitement over the new physics as represented by such figures as de Broglie, Schrödinger, Eddington, Heisenberg, Millikan, Compton, and Einstein. He characterized it as "undergoing a conversion from nineteenth-century materialism to idealism"[26] and saw Jeans's book a fitting sequel to Friedrich Albert Lange's *Geschichte des Materialismus und Kritik seiner Bedeutung in der Gegenwart*,[27] which Kagawa had also translated in 1929 with the help of another colleague.

> Whereas Lange had combated materialism from a Neo-Kantian perspective, we must not ignore the progressive development in the new physics (explored in Jeans's book) that leaves philosophical logic aside and counters materialism on the basis of the content of physics itself.[28]

His confidence that a selection principle in atomic physics had dealt a crushing blow to the old materialism runs like a golden thread through *Cosmic Purpose*:

> The dialectical materialism and materialistic views of history of Marx and Engels were expounded roughly at the same time as Darwin was writing and while atomic physics was still in its infancy. In later years the presence of a selection principle in the material world would be discovered and recognized as governing the atomic realm, something

---

25. *Ibid.*, 50. See also the chart on page 64 detailing the theory of adaptability in the internal environment.

26. Sir James Hopwood Jeans, *The New Background of Science* (New York: Macmillan, 1934). Translated from the English into Japanese by Kagawa Toyohiko and Nakamura Shishio 中村獅雄,『科学の新背景』[The new background of science] (Tokyo: Kōseisha, 1934), Preface.

27. Friedrich Albert Lange, *History of Materialism and Criticism of its Present Importance* (London: Trübner, 1881).

28. Jeans, *The New Background of Science*, Preface.

that did not occur to Marx and Darwin, not to mention the older critical philosophy of Kant.[29]

Kagawa completed *Cosmic Purpose* before the big bang theory had gained widespread acceptance, but one can easily imagine how such a momentous breakthrough would have led him to posit a macrocosmic selection principle corresponding to that of the atomic realm and strengthened his core argument for a directionality that implies purpose.

The final words of Kagawa's Preface to Jeans's book touch again on the spiritual motivations behind his lifelong love of science:

> I have certainly never rejected science. On the contrary, I have stressed that science is itself a window to the spirit. My hope is that through ever more solid work in the laboratory, Japan's natural scientists will break through to the true nature of cosmic reality. In this way, they will come to discover that, in the end, the material universe is an expression of the universal spirit. In anticipation of that day, I offer this book to my Japanese readers.[30]

While Kagawa was critical of scientism, which he saw as an antiquated ideology, these words reveal his strong, and perhaps somewhat naïve, belief that dispassionate scientific research would eventually contribute to a more nuanced, aesthetic, and open-minded interpretation of the purpose of life within the vast reaches of the cosmos.

There can be no doubt that Kagawa considered his extensive reading in the sciences a key aspect of his vocation as a pastor, evangelist, social reformer, and writer. His aim was to help counter the effects of the crass materialism that had cast an ominous shadow over the spiritual life of his rapidly industrializing and militarizing homeland. For him, the clearest impact of this materialism was to be seen in the contemporary face-off between mercantile capitalism and its communist antithesis. In the Preface to *Brotherhood Economics*, a record of the Rauschenbusch Lectures delivered at Colgate-Rochester Divinity School in 1936, he indicates that advances in the natural sciences should also influence economics.

> Theories of relativity and quantum mechanics have completely done away with the nineteenth century's conception of matter and brought unyielding determinism into the realm of possibilities! This means that the twentieth-century materialistic capitalism and materialistic

---

29. *Cosmic Purpose*, 102.
30. Jeans, *The New Background of Science*, Preface.

Editor's Introduction

> communism also must be abandoned. In this book I have tried to open a path to a new social order through a psychological, or conscious, economy.[31]

The connection between physics and economics may not be immediately obvious to readers, but it fits perfectly with Kagawa's "cosmic synthesis" and social vision.[32] That is to say, if capitalism and communism are both grounded in an outmoded materialistic view of the universe and human beings, an economic alternative is needed to complement science's triumph over the old materialism. What makes Kagawa's position so unique and compelling is that he did not stop at utopian pronouncements. While arguing for cooperative economics as a more scientific and humane "third way" that would ameliorate the capitalist greed and communist violence accompanying Japan's modernization, he actually helped found several cooperatives that continue to thrive today.[33] Yet, despite all his practical and theoretical efforts, his hopes of the imminent triumph of this "third way" were dashed in the lead up to World War II and Japan's defeat in 1945.

As already noted, Darwin's theory of natural selection provided the immediate occasion for *Cosmic Purpose*; more specifically, it was the way H. G. Wells had interpreted evolution in his best-selling *The Science of Life* that really drew Kagawa's ire. In the opening pages of *Cosmic Purpose*, Kagawa quotes Wells's comments on the philosophical implications of evolution.

> Variation is at random; selection sifts and guides it, as nearly as possible into the direction prescribed by the particular conditions of environment. Once we realize this, *we must give up any idea that evolution is purposeful*. It is full of apparent purpose; but this is apparent only, it is not real purpose. It is the result of purposeless and random variation sifted by purposeless and automatic selection.... *When we*

---

31. Toyohiko Kagawa, *Brotherhood Economics* (New York: Harper & Brothers, 1936), x.

32. On theological grounds, Reinhold Niebuhr criticized Kagawa's advocacy of cooperativism as a kind of panacea to the world's economic problems: "The forces of evil in the world are really greater than either Dr. Kagawa or his liberal followers believe. And history is a more tragic process than they know." Reinhold Niebuhr, "The Political Confusion of Dr. Kagawa," *Radical Religion* 1/2, Winter 1936, 6–7 (published by the Fellowship of Socialist Christians).

33. For example, the Kagawas founded the Kōbe Co-op in 1920, which today numbers 1.2 million members.

reach man, evolution does in part become purposeful. It has at least the possibility of becoming purposeful, because man is the first product of evolution who has the capacity for long-range purpose, the first to be capable of controlling evolutionary destiny. Human purpose is one of the achievements of evolution.³⁴

In addition to trying to overcome Darwin's materialism, *Cosmic Purpose* is an unrelenting refutation of Wells's conclusions that variation is "random" and evolution is "purposeless." Kagawa points out what he sees as a glaring contradiction in Wells's notion of a "purposeless and automatic selection" that somehow engenders purpose "when we come to human beings":

> We may note that although Wells rejects finality at the beginning of life, he eventually comes to affirm it. How does he account for the contradiction, to which Bertrand Russell also draws our attention in a broader context?³⁵

Kagawa cannot abide this implicit denigration of the long span of time and the many fits and starts—from cosmic dust, atoms, molecules, and ultimately to the advent of myriad forms of life within a supportive environment—that finally made way for the appearance of *Homo sapiens*.

In a later comparison of the merits of mechanistic, vitalistic, and holistic biological theories, Kagawa's exasperation with Wells reaches its height:

> In order to realize the purpose of an individual, an assembly in the form of segmentation is also needed, as is a progressive development.

---

34. *Cosmic Purpose*, 39, from H. G. Wells, Julian Huxley, and G. P. Wells, *The Science of Life: A Summary of Contemporary Knowledge about Life and its Possibilities* (London: Waverly, 1929-1930). (Emphasis added.) A Japanese translation by Ono Shun'ichi 小野俊一 appeared in 24 volumes in 1946 as 『生命の科学』 (Tokyo: Heibonsha).

John Partington comments on the ominous phrase, "controlling human destiny": "Although Wells thus still denies the possibility of creating the perfect human (on the basis that perfection equates to a lack of change inconsistent with evolutionary principles), he nonetheless includes in *The Science of Life* a brief discussion of the possibilities of positive eugenics for the future of humanity." John S. Partington, "H. G. Wells's Eugenic Thinking of the 1930s and 1940s," *Utopian Studies* 14/1 (2003), 75. The connection between evolutionary theory and various theories of genetic superiority would require more extended treatment, especially since Kagawa himself has been accused of voicing such views in his early book, 『貧民心理の研究』 [A study of the psychology of the poor] (1915) and in his statements on the treatment of victims of Hansen's disease.

35. *Cosmic Purpose*, 39.

> Without an appreciation of teleology how can such assemblage and progression be explained? Mechanical views of life deny immanent purpose, neovitalism affirms immanent purpose in each part of the cell, and holistic generational dynamics makes their integration primary. As for those who, in arguing for a holistic biology, deny finality outright, I would reckon them as having fallen back into a neomechanistic position. I find the complete failure of H. G. Wells's *The Science of Life* on this count deplorable. After hundreds or even thousands of descriptions of the amazing adaptability of the biological world, he has the nerve to conclude that it is the result of chance.
>
> Can there be *chance* with *adaptability*? Is there chance that includes sifting and selection? Is chance with law and regulation possible? The logic of Wells is so riddled with contradictions that one can only stand dumbfounded and shocked at its audacity. Those who engage in science need to be a bit more honest. Standing before the marvelous existence of nature, must we not declare our willingness to renounce prejudice and learn humbly?[36]

While Kagawa does not offer explicit reasons for his opposition to Wells, we may surmise that he found his claims "deplorable" for several reasons. In Japan, where spiritual and ethical values center on the ritual expression of gratitude for the benefits of nature and reverence for one's forebears, Wells's claim sounds like an expression of anthropocentric hubris. This may be due to a not so obvious difference in the two thinkers' cultural conceptions of time. If time is generally perceived in modern Europe or North America as a future-directed arrow, thus supporting an "ethic of progress and possibility," in Japan time is generally perceived as an endowment from the past, leading to an "ethic of inheritance and faithful transmission." Reverence for ancestors (*senzo*) and elders (*senpai*), key factors in Japanese religious piety and social ethics, helps engender a strong consciousness of indebtedness to the past.

Given this cultural context, when Kagawa the "scientific mystic" contemplates the appearance of life in the vast span of evolutionary history, he quite naturally extends this "ethic of inheritance" beyond the advent of *Homo sapiens* and all the way back to our ultimate origins in early life forms, proteins, compounds, elements, atoms, and cosmic dust. All these inorganic and organic antecedents are seen as constituents of our embodied existence in the present. This "ethic of inheritance" helps account for the fantastic voyage of Kagawa's book from the atoms to self-conscious-

---

36. *Ibid.*, 165–6.

ness and explain how he was among the first to ponder that the universe makes us what and who we are. When considering the wonder of the evolution of life, Wells opts for blindness over finality, but Kagawa's "logic of finality" holds blindness and finality together:

> I cannot believe events in the natural world are all products of chance. Even if natural events are blind, ultimately they manifest a finality that developed into human consciousness and hence necessitate the examination of the points at which blindness and finality are connected.[37]

When it comes to the advent of life, Wells's assertion of purposelessness is diametrically opposed to Kagawa's conviction of a continuity, which goes further to embrace discontinuity through the finely tuned mechanisms of selection.

At this juncture, we may recall that Kagawa's religious and philosophical reflections were nurtured within the Shinto, Buddhist, and Neo-Confucian *ethos* of his childhood and strengthened by his conversion to Christianity. Unlike other Japanese Protestant leaders who assiduously followed the latest theological trends in Europe (especially Germany) and North America, Kagawa preferred writers like Tolstoy, Ruskin, Bergson, Bowne, and others who resonated more closely with his own religious and philosophical sensibilities. Since such personal convictions bear so heavily on Kagawa's motives for writing *Cosmic Purpose*, it is of interest to note how he expressed these convictions in postwar Japan.

THE POSTWAR MILIEU

Under its Information Bureau and National Mobilization Law, Japan's wartime government exercised strict control over the flow of information. As the situation gradually eased during the US Occupation (1945–1952), a range of formerly suppressed ideas flooded in to fill the national void left by the country's humiliating defeat and loss of common purpose. Because of his wide-reaching international associations and longstanding advocacy of peace, Kagawa had been arrested for questioning by the imperial military police (*kempeitai*) in 1940 and again in 1943. But following the emperor's announcement of surrender on August 14, 1945, he was immediately thrust back into the public spotlight with his proposal to the prime minister's cabinet of a campaign for Total National

---

37. *Ibid.*, 50.

Repentance (*kokumin sōzange* 国民総懺悔)³⁸ and his rapid engagement with forces hoping to rebuild Japan's spiritual and moral life.

As a leader in the Movement for a New Japan, he gave hundreds of talks in the years immediately after the war. In the Preface to a 1947 collection of his postwar speeches, we detect an almost apocalyptic sense of urgency:

> The principle of cosmic repair is nothing other than the expression of boundless love in sacrifice.... Without the social consciousness of collective responsibility and without the tragic, boundless love that repairs and restores sinners, there is no way to recreate humanity. The great love that created the universe and the consciousness that recreates human life is at work in Jesus.... There has never been a more confusing and dangerous time for Japan and the Japanese people than the present. If we do not see that the way to Japan's renewal is in the cross, Japan will degenerate further and further until it is no more than an island in the Pacific Ocean inhabited by barbarians. Only the cross can unify our people and sustain a renewal of culture and scholarship. Yes! The day when the Japanese people discover that the way to the recreation of life is in the cross, Japan will be regenerated and resurrected.³⁹

The mood of national crisis, coupled with Kagawa's commitment to finding renewed purpose in the self-giving, redemptive love of the cross, help us understand why he could not allow Wells's view on "purposelessness" to go unchallenged. There can be little doubt that the core motivation behind all his work was evangelical. Mutō Tomio reports that Kagawa's constant mantra in his postwar lectures, sermons, and interviews was "I must write my theory of cosmic purpose!"⁴⁰ He seems to have been so thoroughly dedicated to the completion of *Cosmic Purpose*, that he even rejected an invitation from a major American publisher to write an autobiography in English.⁴¹

---

38. Robert Schildgen, *Toyohiko Kagawa: Apostle of Love and Social Justice* (Berkeley: Centenary Books, 1988), 248.

39. Kagawa Toyohiko,『宇宙創造と人生再創造』[The creation of the universe and the recreation of human life] (Tokyo: Kamiizumi Shoten, 1947) 1–5.

40. Mutō Tomio 武藤富男,「解説」 [Commentary on *Cosmic Purpose*], *Collected Works of Kagawa Toyohiko*, 13/24: 455.

41. A letter to Kagawa dated April 23, 1954 from Eugene Exman, an editor at Harper & Brothers, reads, "Dear Dr. Kagawa: While I am sorry that you have decided not to write your autobiography, I am pleased to know that you are putting aside time regularly to write your book *The Purpose of the Universe* [here translated as *Cosmic Pur-*

It is against this backdrop that early on in *Cosmic Purpose* Kagawa sets up Wells as his key "denier of finality" and returns to him more than ten times in the course of his argument, marshaling every resource at his disposal from a range of prominent scientists to refute Wells and offer a comprehensive counterargument for purpose.

As Kishi's characterization of Kagawa as "the sole cosmological thinker in Japan" indicates, *Cosmic Purpose* burst on the scene as a complete anomaly. His fellow Protestants were less than enthusiastic. In an otherwise sympathetic 1966 biography of Kagawa, Sumiya Mikio, a well-known professor of economics at Tokyo University and lay Protestant leader, labels *Cosmic Purpose* a "failed experiment." Sumiya concludes:

> Kagawa's is a practical philosophy supported by poetic expression. Thus, while not lacking in conviction, *Cosmic Purpose* may be said to lack scientific argumentation. Despite its copious references to the scientific theories of others, its own theoretical proposals are little more than an interpretation of nature Kagawa had come to on his own.[42]

In hindsight, Sumiya's dismissal of the book has more to say about his own views than about Kagawa's contribution to teleology. To begin with, Kagawa makes no pretense of offering "scientific argumentation." Quite the contrary, in the Preface he describes his approach as a "new, artistic interest in the structure of the universe."[43] Kagawa was reflecting as a philosophically and aesthetically minded religious leader on new developments in science with which he was very familiar, but he did so always and self-consciously as a serious lay reader and not as a specialist.

A less obvious reason for Sumiya's rejection of Kagawa's reading of the scientific data is related to the development of the modern university in Japan. Edwin Reischauer observes:

> The educational system, unlike those of the West, was thus almost

---

pose]. I am eager to talk to you about the manuscript and hope to see you in Evanston. Sincerely, Eugene Exman." Eugene Exman, letter to Kagawa, 23 April 1954, English Correspondence, Kagawa Archives and Resource Center (Tokyo, Japan).

42. Sumiya Mikio 隅谷三喜男, 『賀川豊彦』 [Kagawa Toyohiko] (Tokyo: Iwanami, 2011), 198. Not even Mutō, the editor of Kagawa's collected works, knows quite what to make of *Cosmic Purpose*. "Aside from the question of what contribution this book may make to research, we may call it evidence of the intellect of the kind of figure who appears only once in a century.... It is an intellectual autopsy of Kagawa's brain." *Commentary on "Cosmic Purpose,"* 458.

43. *Cosmic Purpose*, 29.

created *de novo*, and except for some missionary institutions and private universities (which did not compare in prestige with the imperial universities), it was almost entirely in government hands. Thus it was free of the aristocratic and religious domination of many Western educational systems of the time. It was, in fact, a far more rationalized, secular, and state-oriented educational system than existed at that time in most of the West.... It was tailored to meet the national needs foreseen by the leaders: a literate working and military force, a broad group of technicians, and a small leadership elite produced by the imperial universities.[44]

The model of a university based on scientific research had been introduced in Japan as a means of building East Asia's first modern nation state on the foundations of Western learning (*yōsai* 洋才). However, whereas the legacies of the medieval, renaissance, and modern research universities were organically interwoven in the development of European and North American education, by adopting *de novo* the modern Western system in the Meiji Era, Japan made a gradual but decisive break from earlier humanistic forms of learning such as we find it in the academies of Chinese Learning (*Kangaku Juku*).[45] Since Sumiya took the curriculum of the modern research university with its departmentalizing of human knowledge into independent academic specialties for granted, he was not able to excuse, let alone understand, Kagawa's bold incursions into areas thought to be the exclusive reserve of scientists.

Although it is difficult to define the genre of *Cosmic Purpose*, the book is perhaps best classified as a kind of "public natural theology." Given Barth's rejection of natural theology as absolute for all times and places, it is easy to see why Sumiya would have no theological sympathy whatsoever for Kagawa's approach. As baffling as *Cosmic Purpose* has been to Protestant intellectuals like Sumiya, a Roman Catholic scholar like Kishi was able to offer a much more positive reading. *Cosmic Purpose* finds itself more naturally in the company of other works of his time, such as Lecomte du Noüy's *Human Destiny*, Gustaf Strömberg's *The Soul of the*

---

44. Edwin O. Reischauer, *Japan: The Story of a Nation* (New York: Knopf, 1970), 137.

45. See Margaret Mehl, *Private Academies of Chinese Learning in Meiji Japan: The Decline and Transformation of the Kangaku Juku* (Copenhagen: NIAS, 2003). Since Kagawa was among the last generation of Japanese to study as a child in both a *Kangaku Juku* and in the new public educational system, his mental habits were partly shaped by Neo-Confucianism. Kagawa himself acknowledges and affirms this influence. See his 『東洋思想の再吟味』 [A reconsideration of Eastern thought], *Collected Works of Kagawa Toyohiko*, 13/24.

Universe, Fred Hoyle's *The Nature of the Universe*, and George Gamow's *The Creation of the Universe*, all of which Kagawa mentions. What sets Kagawa apart from them, it bears repeating, is that he does not write as a scientist but as a "scientific mystic."

EVOLUTION, GOD, COSMIC WILL

Kagawa was highly critical of Darwin's bias toward philosophical materialism, but he was also an ardent advocate and lifelong student of evolutionary theory. His "logic of finality" in *Cosmic Purpose* testifies to the fact that Darwin's theory helped to shape every facet of his thinking. Already in an earlier book entitled *Love, the Law of Life* (1924), we see him singing the praises of evolution romantically:

> I do not know who first believed in the theory of evolution. It was probably not Darwin or Wallace or Mendel. In accepting evolution we accept more than the mere theories of variation, selection, and the survival of the fittest. Belief in evolution is faith in a progressive entrance into an ever-expanding freedom—from seed to shoot to bud to flower, from anthropoid to human, from man to son of God. What a courageous belief this is! The idea that there is a direct line of evolution from amoeba to man is more daring and romantic than belief in the myth of a Creator making something out of nothing.[46]

As a review of his library demonstrates, Kagawa was indeed much more interested in the latest books on science than in those on theology.[47] And yet, he made no apology for his clear religious convictions:

> Everything that humanity studies—science, economics, politics—belongs to cosmic consciousness. Religion never contradicts science. Science is merely another window for consciousness. To me the theory of evolution has never been contradictory to the idea of creation. Scientists may study the phenomena of life but they cannot create or alter the laws and conditions governing these phenomena. God has given the laws and conditions.[48]

This is not to say that *Cosmic Purpose* can be read as a proof for the

---

46. 「愛の科学」, *Collected Works of Kagawa Toyohiko*, 7. English translation, *Love, the Law of Life* (Chicago: Winston, 1929), 298.

47. 『賀川豊彦文庫』 [The Library of Kagawa Toyohiko] (Tokyo: Meiji Gakuin University, 1963).

48. Bradshaw, *Unconquerable Kagawa*, 103.

existence of God in any traditional sense. In fact, the word *God* only occurs twice in the book, both times in a neutral context. In references to the ultimate mystery behind evolutionary history, Kagawa characteristically opts for the more neutral philosophical term "cosmic will," as we see at the close of his book:

> Besides discovering cosmic purpose, I believe we have to entrust development from here on to an absolute will that has bestowed it with purpose. If that is in fact the case, it makes no sense not to do so. I hold that awakened human consciousness should seek out the support of that transcendent cosmic will in its very struggle to bring everything out into the open.[49]

Kagawa's choice of a philosophical term should not be taken to mean he had abandoned his Christian faith in his final years. It is rather a gesture of respect in addressing a population of whom 1 percent is Christian and whose background is more likely to be Shinto, Buddhist, agnostic, or atheist. His was a "public natural theology" in that it aimed at including the broadest possible readership by focusing on scientific findings of common public interest. By the same token, we may consider the book an example of what Thomas Nagel calls a "natural teleology" that takes the evolutionary emergence of consciousness and reason with utmost seriousness. In his words, "The teleology I want to consider would be an explanation not only of the appearance of physical organisms but of the development of consciousness and ultimately of reason in those organisms."[50]

Kagawa's preference for "cosmic will" over "God" aside, there are no grounds for arguing that he ever let go of his own theological presuppositions. More than any other idea in the history and literature of Christianity, the redemptive love of God actualized in the person and work of Jesus Christ, whose life-giving death signifies the ultimate repair (*shūzen* 修繕) of all things, can be singled out as Kagawa's foundational premise.

Similarly, in spite of the fact Kagawa uses the word "design" over fifty times, and even once uses "intelligence" in relation to "design," the standpoint of *Cosmic Purpose* should not be confused with what is known today as an argument for "intelligent design," at least in its anti-Darwinian forms, or with any theological attempt to refute the theory of evolu-

---

49. *Cosmic Purpose*, 269.
50. Thomas Nagel, *Mind and Cosmos: Why the Materialist Neo-Darwin Conception of Nature is Almost Certainly False* (Oxford: Oxford University Press, 2012), 92.

tion. Quite the contrary, Kagawa finds purposeful design in and through the myriad natural processes that fund evolutionary history. To cite just a few examples of how he discerns "design" in the theories and discoveries of physics, biology, chemistry, and mathematics:

> When he made this discovery (of the periodic table), Mendeleev noticed a wonderful design that reinforced his belief in the truth of the universe.[51]
>
> Chance is also transformed into design by making use of discontinuity in the realm of whole numbers.[52]
>
> ...the universe we inhabit is one of integral geometry and integral mechanics. It is a realm to which purpose-oriented selection has been added and must therefore be thought of as emerging with a "design."[53]
>
> Were atoms not endowed with design, cells would lack the capacity to adapt and to organize the nervous system and sense organs.[54]
>
> Darwin was stimulated by Helmholtz's work on "imperfection in the eye," which for him meant that there is no design or purpose behind the lack of perfection. We are left reading him with a sense of desolation.[55]
>
> Whether we search the inorganic world or inquire into the organic world, we find a design for the appearance of life, whose scale surpasses intelligence and law. One can only bow one's head at the ingenious planning of the universe. The universe appears to be a careless collage of elementary particles, but when we look closely at the atoms that make it up, we cannot but find ourselves up against an *intelligence* hidden in the background.[56]

Where others may see random chance and purposelessness, Kagawa perceives a deep order, complexity, and beauty in and through the natural world—but never in order to refute evolution.

As a refreshing contrast to the endless hostilities between religious and scientific literalists, *Cosmic Purpose* offers neither a religious ideology to trump evolution nor an evolutionary ideology to trump religion.

---

51. *Cosmic Purpose*, 73.
52. *Ibid.*, 83.
53. *Ibid.*, 85.
54. *Ibid.*, 197.
55. *Ibid.*, 214.
56. *Ibid.*, 265.

Rather, it takes contributions of science and religion seriously and sees their insights into the self-conscious wonder of life as complimentary. That there is something rather than nothing, that evidence for directionality points to an as yet undisclosed *telos* in the evolutionary drama— these are facts too large to be encompassed by either religion or science on its own. *Cosmic Purpose* is Kagawa's last endeavor to awaken in readers headed toward a nihilistic materialism a renewed sense of awe before life and consciousness in the immensity of this universe in the making.

All of this clears him of the suspicion of engaging in unscientific arguments to deny evolution. Indeed, Kagawa wants to be taken seriously as a modern Christian thinker who, for example, rejects astrology and "all superstition regarding numbers." In this connection, he agrees with Arrenhius's rejection of Tycho Brahe's kind of "natural theology" and uses it to lead into his own approach to finality.

> The primary mistake here is understanding this kind of astrology as purposive. It seems to me superstitious to ascribe to the extreme arguments advanced by Tycho, who went too far in mixing up the laws of physics with those of physiology and psychology.
>
> The affirmation of finality we have in mind here is of a different sort. I will insist throughout on a developmental process in which the principles of selection in nature, combined through organic chemistry, make life possible. And I will further insist that it is this world of life that, in turn, makes up a purposive world through the organization of the nervous system.[57]

As for numbers, he has this to say:

> The ancient Egyptian philosopher of mathematics Ptolemy tried to ground philosophical principles in numbers. Strong hints of his influence are present in Plato and Aristotle. I would like to recall those traces and then take the further step of obliterating all superstition surrounding numbers so that, grounded in the image of the universe we have received from modern physics, we may discover a new standard for integral ratios from the viewpoint of the principle of natural selection.[58]

Kagawa seeks to base his own argument for purpose purely on findings in modern physics, biology, mathematics, and chemistry.

57. *Ibid.*, 41.
58. *Ibid.*, 66.

OVERCOMING ANTINOMIES

Cosmic Purpose pays a great deal of attention to overcoming what Kagawa calls the Kantian "cosmology of antinomies"[59] between determinism and free will, matter and spirit, and chance and purpose as a basis for offering a view of "cosmic evil." On one hand, his commitment to a "cosmic synthesis" allows him to bind the antinomies together in a complementary relationality, reflecting similar Neo-Confucian approaches to the dialectic between *li* (principle) and *ch'i* (material force).[60] On the other, by always stressing the limiting function of a "field" within which freedom, spirit, or purpose operates, he clearly assigns a marginal control to the field. It would seem that, despite his best efforts, he cannot avoid mechanistic determinism altogether. Needless to say, many scientists and philosophers today take a similar position in limiting free will, spirit, or purpose to a small field of operation within a mostly determined universe.

On the hotly debated subject of determinism versus free will, Kagawa makes only one theological statement and it seems to tip the scales in favor of determinism: "Theologically, Calvinistic determinism and Arminian free will theory can be harmonized if seen as the free will of variation within a field."[61] This statement clearly favors the Calvinist side of the debate. Hence, in spite of his lack of patience with theological debates, he remained to the end a Presbyterian pastor under the sway of the classical Reformed theology he had studied in Kōbe and Princeton. To debilitate will or reduce it to a variation within a determined field would result in a certain loss of freedom. However the freedom is circumscribed and for whatever reason, it ends up less than freedom. Taken together with

---

59. *Ibid.*, 199.
60. In Japan, on the one side of this Neo-Confucian development were those who stressed *li* 理, represented by the Shushi (Zhu Xi) School, and on the other side were those who stressed *ch'i* 気, represented by the Ōyōmei (Wang Yangming) School. Whereas the first *Shushi* School tended toward a rationalist view of ultimacy, the *Yōmei* School preferred intuitive or mystical perception. See Yamashita Ryūji, "Nakae Tōju's Religious Thought and Its Relation to 'Jitsugaku'", *Principle and Practicality: Essay in Neo-Confucianism and Practical Learning*, Wm. Theodore de Bary and Irene Bloom, eds. (New York: Columbia University, 1979), 307–35. Kagawa had deep respect for Nakae Tōju (1608–1648), the founder of the *Ōyōmei* school in Japan. See Kagawa Toyohiko,『東洋思想の再吟味』 [A reconsideration of Eastern thought], *Collected Works of Kagawa Toyohiko*, 13/24: 132–5.
61. *Cosmic Purpose*, 243.

certain tacit Taoist, Buddhist, and Neo-Confucian tendencies in his other writings, Kagawa's Calvinism cannot avoid the impression of a certain mechanistic determinism that does just that.

While seeking a vitalistic view of mechanism and purpose that prepares for and is enhanced by life, he describes the subtle working of mechanisms, from the atomic level up through chemistry, biology, and psychology, that embrace a wide range of contingency. By finally limiting freedom, however, he seems prematurely to impose the working of a certain *deus ex machina* on the yet to be disclosed ultimate *telos* of the evolutionary drama. We would at least expect Kagawa to allow for the possibility that future research in astrophysics, subatomic physics, chemistry, or neuroscience may reveal the existence of a hitherto inconceivable degree of freedom that may moderate if not overturn deterministic perspectives.

Kagawa's view of "cosmic evil" operates in a similar way as "slippage" (*zure*)[62] or gaps in the "field" of natural selection that ultimately serve some higher integration or purpose. For a book that claims to be a reconsideration of "the problem of cosmic evil from the perspective of cosmic purpose," Kagawa has surprisingly little to say about the matter. From a Christian theological perspective, his remarks represent an inadequate description of the evil or nothingness (*das Nichtige*) that Barth, for example, describes as a kind of threat to God and by extension to humanity, a chaos or absurdity that has no ontological status of its own but which is nonetheless real as a kind of "impossible possibility." According to Barth, only the uncreated One, and not any selective purposive mechanism within the natural order, can deal with *das Nichtige*.[63] In the person of Jesus Christ, God has made just that gracious decision to free creation from futility.[64] Here, Kagawa's lifelong attention to the reparative redemptive love of Jesus Christ calls for more adequate theological treatment.

Having transported readers back and forth between the microcosm and the macrocosm several times on what may only be described as a comprehensive quest for purpose within us, around us, and beyond us, we are left with the question: *When all is said and done, what does Kagawa see as the ultimate purpose of the cosmos?* His final answer is to

---

62. Ibid., 213.

63. Karl Barth, "God and Nothingness," *Church Dogmatics*, III/3 (Edinburgh: T. & T. Clark, 1961) 289–368.

64. Rom 8: 19–21.

align himself with Dante in mystical silence before "the impenetrability of cosmic purpose":

> In his *Divine Comedy*, the medieval Italian genius Dante resigned himself to our inability to answer the question of the final purpose of the cosmos....
>
>> For that which you ask lies so deep within the abyss of the eternal statute that it is cut off from every created vision. And when you return to the mortal world, carry this back, so that it may no longer presume to move its feet toward so great a goal. The mind, which shines here, on earth is smoky, and therefore think how it can do below that which it cannot do, though heaven raise it to itself. (*Paradiso*, canto 21)
>
> Such was Dante's surrender to the impenetrability of cosmic purpose, and I find I must agree with him.[65]

And so Kagawa joins the contemplatives in the Seventh Heaven, where the long sought after heavenly vision is finally "cut off from every created vision."

The horizon against which Kagawa painted his religious-aesthetic vision is as wide as the cosmos itself, but only that wide. On the possibility of a "cosmic will" behind or above the cosmos, he remains silent. Theological difficulties that arise when divine freedom and will are set against the presence of the *imago Dei* in the created world do not seem to have troubled him.

While one is awestruck at the original and thoroughgoing attempt of *Cosmic Purpose* to discern purpose at every point in the evolutionary history, it is hard finally to agree with Kishi that the book offers an "evolutionary doctrine of creation." For Kagawa, theological reflection is ultimately sacrificed on the altar of mystical vision. Consequently, the lack of substantive reference to the Creator disqualifies it as a contribution to the theology of creation, yet we should also recall that Kagawa seems to have avoided explicit theological language intentionally in order to make a broader public appeal for this theory of cosmic purpose.

What, then, has this scientific mystic's "public natural theology" or "natural teleology" to contribute to the dialogue between science and religion? Kagawa himself had hoped to lay to rest once and for all the philosophical materialism he saw behind Darwin's theory of evolution,

---

65. *Cosmic Purpose.*, 267–8.

and to elaborate a universal "logic of finality" to overturn Wells's views on "purposelessness" and "chance." Obviously his hopes were not realized on either count. Nevertheless, in making *Cosmic Purpose* available in translation for the first time, our hope is to add a distinctive Japanese voice to the ongoing search for positive alternatives to radical materialism and its view of evolutionary history as nothing more than the purposeless workings of random chance. As Kagawa's last will and testament, *Cosmic Purpose* challenges those in religion, theology, and natural science to deeper public engagement on critical philosophical issues facing humanity. The science and religion conversation awaits a new generation of creative thinkers who will take up Kagawa's challenge and further his quest for a "cosmic synthesis."

Acknowledgements

This translation represents a truly international collaboration between Japan, the United States, and the United Kingdom. To begin with, I would like to express my sincere gratitude to Kayama Hisao, Director of the Kagawa Archives and Resource Center in Tokyo, Inagaki Hisakazu, Professor at Tokyo Christian University, and Kagawa Tokuaki, Director of the Kagawa Memorial Hall in Kōbe. They and their colleagues have provided invaluable advice and practical support for this project from its inception. I also wish to thank David Vikner, President of the Japan ICU Foundation, and his staff who have welcomed my research as a contribution to their own mission. I want to offer a special thanks to Conor Cunningham of Nottingham University and Editor of the Veritas series. From our conversations in Princeton, Conor was able to recognize the value of Kagawa's contribution to ongoing debates on the question of teleology. Finally, I wish to thank my translator, James W. Heisig of the Nanzan Institute for Religion and Culture. Truly, Jim may have been the only scholar in the world capable of taking on such a colossal task, and he did it with characteristic grace, humor, and alacrity. As editor, I take full responsibility for any errors that remain.

I also want to express my personal thanks to Carol, our four children, Rose, Paul, Sarah, and Katie, my mother-in-law Jo Tolley, our extended family, and several close friends and colleagues, Chiba Shin, Joseph Dunkle, Claudia Genung-Yamamoto, Barry Jacobs, Paul Johnson, Mark Mullins, Haruko Nawata Ward, Richard Osmer, Torigai Keiyō, and Wentzel van Huyssteen.

I dedicate this book to the family of Kagawa Toyohiko and Kagawa Haru and the many people in Japan and around the world who have been touched by their enduring witness.

Preface

I began to wrestle with the problem of cosmic evil at the age of nineteen. Around 1912 I called on Dr. Mizuno Toshinojō of Kyoto University to inquire about studying atomic theory. In July of 1914, with the onset of World War I, I headed for Princeton University where I majored in mammalian evolution. After that I stole some time from my preoccupation with social movements in Japan to continue my study of "the universe and its salvation." At the time of the China incident the secret police put me in solitary confinement in Shibuya, Tokyo, for my involvement with the peace movement. I was subsequently transferred to Sugamo Prison, where I passed my time reading books on the evolution of mammalian skeletons.

Shortly before the outbreak of the Pacific War, I began to reconsider the problem of cosmic evil from the perspective of cosmic purpose, which brought me to a new, artistic interest in the structure of the universe. I had a deep sense that the mysterious unfolding of the structure of the cosmos was still in process. Without rushing to any conclusions, I felt the need to focus on the grand production of the universe. But as it is not enough for these all too solitary thoughts of mine to reach beyond a small circle of acquaintances, I would like to make public portions of my view of the universe and appreciation of its artistry.

I am grateful to Yoshimoto Kōzō for his help with the proofreading when problems with vision made the task difficult for me. Nakamura Hiroshi offered advice on the food choice of microorganisms and Kuroda Shirō organized drafts of the manuscript for me. Foreign scientists to whom I am indebted include several Nobel laureates, especially Robert A. Millikan, Arthur H. Compton, Linus Pauling, and Ernest O. Lawrence, who guided my inexperience in experiments and responded to my doubts one by one. Finally, I would like to express my thanks again to Honda Chikao, president of the Mainichi Newspapers Co. for seeing this book to publication.

Kagawa Toyohiko
Summer, 1958

Directionality in Natural Selection

1 Natural Selection and Directionality

The universe as a giant chess board

The universe is a giant chess board. The hundred and more pieces we call atoms, like the pieces used in Japanese chess, follow a set of rules known as atomic values, whose movements differ in valence from 1 to 8. On learning of this, I came up with the idea of "atomic chess," which I then played with some friends. Hydrogen would be the "king," whose capture would spell defeat. The chess board we used measured 18 rows by 18 columns, like a Japanese *Go* game board. The pieces were positioned following Niels Bohr's arrangement of the periodic table with seven rows from the bottom carefully arranged in order: 2–8–8–18–18–32–6. Since neither the six new elements discovered in the United States during the Pacific War nor the three atoms recently discovered were known at the time, the front line was comprised of six radioactive pieces.

When I was arrested by the secret police in 1940 for my participation in the peace movement, I was sent for eighteen days to a solitary cell, stifling with the late summer's heat and infested with mosquitos. It was there in my cell that I came up with the idea of atomic chess.

Assuming that the size of the diameter of atoms needed for crystallography had been learned in primary school, my aim was to try to make the atomic chess pieces the equivalent of one hundred million angstroms (an angstrom being one hundred-millionth of a centimeter.) As this proved too expensive, I divided the pieces into thirteen forms roughly approximating the required size.

I found it interesting that school children familiar with Japanese chess were all able to play atomic chess. Alkaline soils with a valence of 1 were assigned the same movements as a "pawn." Elements with a valence of 2 like beryllium, magnesium, calcium, zinc, strontium, cadmium, barium, mercury, and radium, were like the "lance," except that instead of moving only one way they were allowed to move forwards and backwards.

Elements with a valence of 3 could move in three directions; those with a valence of 4 could move forwards and backwards as well as to the right and left; those with a valence of 5 could move in five directions like the "silver general"; those with a valence of 6 could move in six directions like the "gold general." Elements with a valence of 7 were assigned the movements of the "king," except that they could not go backwards; those with a valence of 8 enjoyed all eight directions of movement permitted the "king": forwards and backwards, left and right, and diagonally forward backward both left and right. Radioactive elements were given the powers of the "rook" and "bishop." Only they were allowed to capture the non-compound atoms of helium, neon, argon, krypton, xenon, and radon. Following the textbook rules of chemistry, negative (minus) pieces were able to capture the positive (plus) pieces, and the first side to have its hydrogen atom captured would be the loser.

Selection and the aim of play

Games that do not end with a winner and a loser are uninteresting. Winning and losing involves choices. *Sumō* wrestling, baseball, and football all have referees or umpires who make clear choices. Choices that have an affect, even the simplest of choices, are what makes it a game. Whether it is a game of tag or a game of marbles or even just a flip of the coin, the fun lies in those choices and the struggle for the advantage.

We may have our doubts about what the purpose of life is for lower animals, but just as our children spend a day playing and having fun, it seems to me that lower animals, too, with no more than their instincts to go on, make a game of choosing among their primitive instincts to enjoy the short life that is given to them.

I once visited Dr. Ishikawa Chiyomatsu, now deceased but then director of the Seaweed Institute at Hokkaidō University located off Muroran Bay. He showed me under a microscope the actual reproduction process of seaweed. Because it happened to be early May, the season when seaweed reproduces, it was possible to view under a microscope the spectacle of a seaweed sperm approaching an ovum. The moment the two sexes approach one another the sperm begins to dance around in a circle, rotating like an airplane propeller. I am unaware of what biological significance this has, but I was unable to contain myself for wonder at why this surprisingly energetic movement was taking place.

I marveled at the excitement involved in the activity of selection. If interest in selection were not tied to the survival of the species through sex, the species would be in peril of extinction. This excitement at the time of selection is visible in all lower life forms, particularly in animals. I do not consider this merely a matter of sex. Consider the excited dance of the honey bee at having discovered food. Each and every choice made in the course of its instinctual life is a joy, and it is in the succession of those joys that the bee goes on living.

Philosophers tell us that when all is said and done, even the economist absorbed in seemingly boring commercial exchanges has not taken a single step further than seaweed and the honey bees in the matter of selection. The lady of leisure killing time at a department store knows the joy of choosing, "Shall I buy this one? Shall I buy that one?" The mentality of the postwar generation that has taken to bicycle races and horses races, or of the newspapers and weeklies that write up quiz programs and *Go* and chess matches as if they were world events, is not in the slightest removed from the tendencies apparent in the selective movements of the weeds and the bees. The presence of such tendencies in the process of selection poses an interesting question.

A polynomial, high-level compatibility equation

An airplane in a dense fog is guided by radar to a landing. Similarly, breaking through the thick and chaotic fog of the material world, the pulse of life dispatches mammals with consciousness into the world of the spirit. Its guidance is far more admirable than radio radar. Early humans took this up in religious forms with belief in the existence of supernatural forces.

It is only relatively recently in the history of humanity that doubts were raised about the presence of a supernatural guiding force deep within the material world. Humans invented "machines," and with that the cosmos came to be thought of as a machine. Humans also arrived at the idea of "chance," and with that the universe and human beings came to be thought of as chance events. Of course there are mechanical things in the universe just as there are realms of chance. Further, the actions of human beings in the universe are also subject to its finality. Even in the world of atomic physics, which prides itself on being the most scientific and mathematical of disciplines, the so-called "selection principle" has been universally accepted.

In the age of Immanuel Kant, no selection principle was permitted control over the material world. Indeed there is not a single line in his *Critique of Pure Reason* even touching on the question. Kant's age was dominated by a mechanical, Newtonian view of the universe that did not so much as notice the presence of a principle of selection driven by inclination. With the advance of chemical spectroscopy, however, the world of ultrashort waves drew open the curtains on the human intellect and brought conclusive evidence of the existence of radio radar that can shed light into the blind alleys of our humanity. The macrocosmos contains an *adaptation to the natural environment* that has given birth to life, and it turns out that elements that keep silent like so much idiotic matter are not as foolish as we have thought. It has become clear that each of the elements has its own selectivity and that each element's energy is committed to a certain orientation; that chemical compounds have a disposition to arrangement; and that there is tendency and directionality in the generation of things.

The universe that equipped life to accommodate to the natural environment also prepared the inner environment of organisms for adaptation, and they did so in order to invest life with adventure. For an organism born in the sea, for example, the ascent to dry land represented an extraordinary adventure.

Thus the capacity for outer adaptation came to manifest an adaptability in the inner environment of living beings—blood, for instance—that is nothing short of astonishing. This congruence of outer and inner meant that an organism has taken on the ability to direct the protection and defense of its life form.

For life, selective adaptation did not stop there. In stepping out of the sea and onto dry land, life had to prepare for itself a bodily form with the mobility needed to fly through the sky.

The grand design to preserve energy that succeeded in using minute standing waves of light selectively to create plus and minus electrons also led to the invention of a surprising structural adaptation in life activity by way of the oft-cited metallurgy phase rule or "degree of freedom" advanced by J. Willard Gibbs. In a stable universe, solids, liquids, gases, heat, electricity, magnetism, and radiation exist separately, each in its own place and each with its own movements. Not so with living organisms. At a suitable temperature the liquids, gases and heat, electricity, magnetism, and radiation that began to be galvanized through combinations of triphosphates and proteins, like adenosine triphosphate, and to be physically isolated, came to demonstrate an extreme adaptability in both location and movement, leading to the surprising activity of the living organism.

Not satisfied with this kind of high-level adaptation, the selectivity of the universe took a further step in adaptation, synthesizing the experiences of two genealogically distinct organisms through the arrival of the sexual function. The organizational design of the two sexes was not aimed at merely living; it included an ingenious scheme for bringing about a better life. Just as one party can reach another by dialing numbers intermittently on the telephone, it is as if the sexual function were being summoned at some point in the order of the evolution of life and in accord with a cosmic arrangement carried out by the heat and radiation of the sun. The delicate activity of egg and sperm show adaptation. A few exceptions aside, even the lowest form of plankton is invariably engaged in sexual activity. It is through sexual activity that life was organized for higher and better advancement.

The way to this higher advance was accelerated by a divergent evolution in which male and female characteristics were carried genetically. The mitogenesis of sexual cells expresses this adaptation wonderfully. It carries out additions and subtractions, multiplications and divisions of chromosomes that would elude the planning of even the most skilled engineer. Furthermore, the method for preserving the "species" adds a degree of adaptive selection to the world of living things. The mutual nutritional functions in play among living things exhibit the struggle for survival needed for "balance in nature": on the one hand, the transformation of the inorganic into the organic in plant life, and on the other, the sacrifice of lower animals that feed off plants as food for higher animals that enjoy a form of life with greater freedom.

This polynomial, high-level compatibility equation with its manifold structure aroused adaptation in the sensory instincts of living organisms, both in terms of acute adjustments in physical chemistry as well as of adaptations with a high level of physiological and psychological purpose. Every kind of animal instinct, unconscious and semiconscious, wakens physiological desires at a psychological level. Birds fly from the North Pole to the South; fish follow temperature and food around the oceans.

In this way the awakened psychology of living organisms gave shape to a great eruption in nature: creative, purposive, conscious will.

Life born and reared in the world of nature now comes to birth as *mind* with its own finality and creative will. May we not conclude that the finality of mind is oriented in a "cosmic direction"? As I see it, the mechanism embedded in the universe is structured like a polynomial, high-level adaptability such that in addition to affirming the genesis of purpose-oriented mind as the culmination of adaptation, we may attribute to mind the essence of finality at the ground of the universe itself.

Deniers of finality

Many of those who possess a materialist view of the universe reject arguments for finality. Georgiĭ Fedorovich Aleksandrov, a fellow of the Institute of Philosophy of the Russian Academy of Sciences, is one of them:

> Adaptation is useless as proof of the argument for purpose. The argument for purpose is an enemy of science. It sees human beings as the masters of nature, but in discovering the true connections between things, it blocked the way to our knowledge of those connections.[1]

H. G. Wells, co-author of *The Science of Life,* rejects purpose. He even claims that the words of Shakespeare could be written by accident. He writes:

> Variation is at random; selection sifts and guides it, as nearly as possible into the direction prescribed by the particular conditions of environment. Once we realize this, we must give up any idea that evolution is purposeful. It is full of apparent purpose; but this is apparent only, it is not real purpose. It is the result of purposeless and random variation sifted by purposeless and automatic selection.
>
> The term purpose has a very definite meaning. It is a psychological term, describing a certain familiar state of our own consciousness: it implies the prevision of an end, and a determination to reach that end. For evolution to be purposeful, one of two things must be true. Either living things themselves must be purposive in their evolutionary changes—the flower must somewhat want to attract the bees, the horse intend to lose all its toes but one; or else, although the living animals and plants themselves betray no purpose, purpose must exist in the mind of a divine Being, who is manipulating life and its environment to bring out His purpose, as we manipulate matter and events to bring about our purposes.
>
> The first alternative is that of Bergson and Shaw; as we have seen, we must dismiss it. Variation and selection in themselves are blind….
>
> Evolution has deprived barnacles and oysters of movement and brain; it has produced the female mantis, who begins eating her mate

---

1. 『弁証法唯物論』 [Dialectical materialism], trans. by Kozai Yoshishige 古在由重 and Mori Kōichi 森宏一 (Tokyo: Aoki Shobō, 1955), vol. 1, 45. [As this work is not available in English, the passage was rendered through the Japanese translation.]

during the act of pairing; it has generated the bloodthirsty land-leeches and mosquitoes, and fitted the ichneumon-fly grub to devour its living caterpillar prey slowly, from the inside; it has brought into being not only strong, intelligent, and beautiful creatures, but also degenerate parasites and loathsome diseases....

When we reach man, evolution does in part become purposeful. It has at least the possibility of becoming purposeful, because man is the first product of evolution who has the capacity for long-range purpose, the first to be capable of controlling evolutionary destiny. Human purpose is one of the achievements of evolution.

Human purpose has arisen as a product of the mechanical workings of variation and selection. But now that consciousness has awakened in life, it has at least become possible to hope for a speedier and less wasteful method of evolution, a method based on foresight and deliberate planning instead of the old, slow method of blind struggle and blind selection.[2]

We may note that although Wells rejects finality at the beginning of life, eventually he affirms it. How does he account for the contradiction, to which Bertrand Russell also draws our attention in a broader context?[3]

Arrhenius' misunderstanding of finality in science

Throughout his historical study of the theory of creation, the Swedish scientist Svante Arrhenius resists views of the universe that claim finality. He considers astronomical ideas supporting such claims to be astrological in character. To be sure, the mixing up of physics and psychology in assertions that "this star and that have an influence on each and every one of us" is something sixteenth-century astronomers were ready to accept. Arrhenius cites Tycho Brahe to make his point:

> Those who deny the influence of the stars also resist divine omniscience and providence; they further reject what is a matter of the clearest experience. The idea the heavens adorned with these brilliant stars and so astonishing in its scale and precision were made by God to no purpose at all is simply absurd. Even the actions of the biggest imbecile are not without a purpose. Opponents may reply that the

---

2. H. G. Wells, Julian Huxley, and G. P. Wells, *The Science of Life: A Summary of Contemporary Knowledge about Life and its Possibilities* (London: Waverly, 1929–1930), 428.

3. See the chapter on "Cosmic Purpose" in Bertrand Russell, *Religion and Science* (London: Thornton Butterworth, 1935), 190–222. [It may be noted that Kagawa took over the Japanese translation of the chapter's title as the title of the present book.]

heavens are no more than a clock reporting the days and months and years. But if that were all, the sun and the moon would already be more than enough. For what possible purpose might the other five planets be moving about in their own spheres? Slow, plodding Saturn takes thirty years to complete a cycle; Jupiter with its flickering rays rounds its orbit once in every twelve years. Mars and Mercury need only two years, but why are they there? And why others, like that beautiful Venus who, as maidservant of the sun, glistens now as the evening star and now as the morning star? Beyond these, what reason is there for the eight other stellar constellations in the celestial realm? Nor should we forget the smallest among the countless stars, the smallest of which is nearly a thousand times that of the earth and the larger of which are a hundred times greater still.

.....

Experience shows that the celestial bodies each exert their own power over the earth, like the sun which generates the cycle of the seasons. In tune with the waxing and waning of the sun, the gray matter and bones of the animals, the pith of the trees, the flesh of the crabs and snails grow and decay. The sun has an irresistible power that lifts up the waves of the evening and morning tides, so that they swell when it prompts them to and ebb when it weakens its hold. When Mars and Venus meet, the rains fall; when Jupiter and Mercury come together, they bring wind and rain and lightning. And if these wandering stars should appear together with a fixed star, their force becomes all the greater. When a planet that brings moisture meets a damp constellation, long rains result. If a dry planet groups with a hot constellation, a period of terrible drought follows. These are all things we know well from everyday experience. The heavy rains of 1542 coincided with a remarkable coming together of planets in the constellation Pisces. In 1540 first there was a solar eclipse in Taurus and then a confluence of Saturn and Mars in Libra, which was followed by a confluence of the sun and Jupiter in Leo. That summer was unusually sweltering. Or again, who can forget the impact when, in 1563, Saturn and Jupiter met with Leo and drew close to the faint stars of Cancer! Already of old, Ptolemy saw that these stars can suffocate people and cause epidemics, and indeed this is precisely what ensued during the following years when thousands of people in Europe were sent to their graves by the plague that raged.[4]

---

4. Arrhenius アーレニウス, 『史的に見たる科学的宇宙観の変遷』 (Tokyo: Iwanami Bunko, 1951), 114–5. [The lengthy quotations from Tycho are from the Japanese; they were omitted from the English translation: Svante Arrhenius, *The Life of the Universe as Conceived by Man from the Earliest Ages to the Present Times* (New York: Harper & Brothers, 1909), 96–7.]

Natural Selection and Directionality

The primary mistake here is understanding this kind of astrology as purposive. It seems to me superstitious to subscribe to the extreme arguments advanced by Tycho, who went too far in mixing up the laws of physics with those of physiology and psychology.

The affirmation of finality we have in mind here is of a different sort. I will insist throughout on a developmental process in which the principles of selection in nature, combined through organic chemistry, make life possible. And I will further insist that it is this world of life that, in turn, makes up a purposive world through the organization of the nervous system.

The Principle of Natural Selection

Taking hold of a mosquito net

When the time comes to hang a mosquito net for the summer, I first take it out of the cupboard where I had left it folded up. I was always finding myself at a loss how to begin so that I didn't get the long and short ends mixed up, so I marked the flap on one of the lower corners in red and everything went smoothly after that.

What happens when a plant puts forth buds or a chicken egg begins to hatch is a very complicated process and requires skillful judgment to see it through. For things like chicken eggs, judgment proceeds beautifully and without a hitch for a mere twenty-one days. We humans may get confused in deciding how to pull open the four flaps on a mosquito net, but when we look at the development of the unborn, unconscious egg and the miraculously good judgment it exercises for an animal with life and instinct, with a mechanism of development and a self-determining purpose, we cannot help but notice within the shell a certain selective activity with an amazing directionality (judgment) to it.

This is true not only of organic beings; one notices selection at work in the discriminating movement and aggregational activities of the inorganic world as well. As Lawrence Henderson explains, when cosmic life is put at the center, it is hard not to believe that the inorganic and organic activities of physical chemistry are directed towards the manifestation of life. This means that there is direction to the production of matter and life. But even if one is not prepared to go along with that conclusion, there is no denying the fact that such phenomena and processes exist.

In terms of the quantity and quality of freedom to act, life has a far wider reach than that of atoms and molecules. If we grant that the index of progress in "freedom" is the ability to use the smallest amount of energy to produce the greatest amount of activity, nothing approaches the high degree of freedom found in life systems. Naturally, the structures of the psyche raises the level still higher. Even in terms of energy efficiency, we may consider the orientation of matter as directed towards life. As has been noted earlier, this orientation is not blind; it is through and through a matter of selection.

Fate, chance, and the meaning of selection

Nevertheless, we find two opposing tendencies from the earliest beginnings of science: to treat things as matters of fate or to treat them as matters of chance. Hardly anyone has adopted the selection principle as a law governing the development of the universe. Up until around the turn of the twentieth century, mechanistic theories based on laws were dominant; then theories of chance and probability came to the fore. Nevertheless, as Lecomte du Noüy stressed, theories of chance alone cannot account for the emergence of the mechanism of proteins.

The arrival of the selection principle held out the possibility of overturning theories of mechanistic fate and countering theories of blind chance with a fixed system of laws and restrictions:

1. Selection does not occur where there is no change.
2. This further entails the existence of standards.
3. These standards take the form of an a priori probability in the inanimate world, but in the animate world, organisms make choices based on their own internal normative standards. The choice of mates is of this sort.
4. Therefore, the selection principle functions as a kind of mediator between *change* and *development*.
5. In temporal terms, the selection principle points to a tendency to progress in a certain direction and make adjustments.
6. The selection principle shows restrictions limiting change related to space.
7. In the realm of selection, the laws of cause and effect imply a selective closure. For example, A does not necessarily become A' but follows its inherent degree of freedom.
8. Paradoxically, where there is no choice, there is no concept of *time*.
9. Where the selection principle does not come into play, there is no need to introduce the function of *field*.
10. Accordingly, the condition of working in *space and time* means

that selection entails the continual guidance and restriction of things with a defined tendency.

The evolution of sexuality and the emergence of selection

In trying to analyze the evolution of *sexuality*, we see that the selection involved is one of a priori probability. The philosopher Arthur Schopenhauer argued that the establishment of sexuality points to an a priori content determined by cosmic will. Males and females are sexually determined prior to their awareness of sexuality and are designed to feel instinctive impulses with a finality to them. If sexuality is considered to be blind, then sexual impulses might also be blind. But if we look at things in terms of how sexuality distinguishes male from female and gives birth to a purposive world in which they seek each other out, we may say that an impulse-driven finality is latent in selection. At the very least we have to confirm with Aristotle and Kant the presence of a latent organic purpose.

Why did selection emerge in the first place? Obviously, within the world of living things selection is a requirement of living things as such. But as Arnold Sommerfeld has pointed out, when we come to the activity of electrons within the inner orbit of the atom, we find an unmistakable principle of selection at work. Given that selection is traceable in the world of physics as a matter of a priori probability, we are obliged to acknowledge a selectivity latent in the universe.

The structure of selection is furnished complete with the elements of energy, change, and development. For these elements to be assembled, the mechanism of a "field" must also be provided. Since each element creates a node *in order to* construct a selection, we cannot but acknowledge the possibility of selection opening out into a teleological realm.

Despite the insistence of Darwinians that natural selection occurs by chance, the a priori selection of physics cannot be considered a matter of chance. How would we have an eyeball if everything were thrown together blindly? Why would we have reason rather than contradiction?

If you were to present uneducated people one of Luther Burbank's improved "seedless prunes," they would probably think it was just one of those things that happen from time to time. But if the production of a seedless prune were explained to them in terms of the selection principles of thremmatology or scientific plant breeding, they might discover that something more than chance was in control of the prune.

The emergence of selection marks the *beginning* of something in which selection commits itself, as it were, to change and decision. Selection becomes all the more complex with the advance from physical to chemical selection, which then further develops into the world of life. The phenomenon of life cannot come about without biochemical selection being added to the whole of physiological selection and then biological selection added on top of that. Following this line of thought, the phenomenon of life may seem to be a kind of subtraction by selection, but it is nothing of the sort. On the contrary, the calculus of adding, subtracting, multiplying, and dividing involved in selection occurs at high speed and takes the form of an integrated process involving multidimensional planning.

A great deal of physical selection involves a priori probability, much of which is oriented to mainly mechanical decisions. If we restrict ourselves just to physical selection, the complex of latent decisions can be seen to cover, among other things: (1) direction, (2) position, (3) axis, (4) coordinates, (5) fulcrum, (6) priority, (7) surface, (8) angle, (9) tendency, (10) numerical value, (11) ratio, (12) refractive index, (13) magnetic field, (14) direction of magnetism, (15) electricity, (16) electrical direction, (17) energy level, (18) atomic valence, (19) bonding direction of atomic valence, (20) configuration, (21) periodicity, (22) frequency, (23) wave length, (24) amplitude, (25) hydrogen ion concentration, (26) velocity, (27) permutation, (28) organization, (29) series, (30) temperature, (31) humidity, (32) density, (33) coefficient of thermal conduction, (34) melting point, (35) evaporation point, (36) , and (37) directional angle of atomic valence.[5]

When it comes to life, not even the selections of energy are simple. Atomic energy, electrical energy, gravity, pressure, hydraulic pressure, solubility, evaporation, crystallization, magnetism, attraction, repulsion, and so forth—all are contained in the phenomenon of life.

Chemically speaking, things are still more complex. Life would not have appeared without a concentration of elements that depends on the selective integration and combination of their special characteristics. For "life" to appear on earth, countless selections had to be brought together in proteins, fats, starches, vitamins, hormones, and the like.

---

5. Linus Pauling, *The Nature of the Chemical Bond and the Structure of Molecules and Crystals: An Introduction to Modern Structural Chemistry* (Ithaca, NY: Cornell University Press, 1940), ch. 4.

Forms of selection

Selection takes a variety of forms. Michael Faraday discovered that when a needle is brought close to a ring of magnetized copper wire through which an electric current is passing, the needle tends to turn in a direction at right angles to the wire, whereas when the current is reversed, the needle turns in the opposite direction. We may refer to this uncomplicated display of directionality that occurs according to simple a priori probability as *simple selection.*

When it comes to a complex life phenomenon such as the cell, even in the calculus of automatic selection something truly astonishing happens:

1. Simple selection (many phenomena of physical selection)
2. Dual selection (harmonious selection), as in plus and minus electrical charges
3. Complex selection (complicated harmonious selection)
   A. Correlative selection
   B. Proportional selection
   C. Substitute selection
   D. Differential selection
   E. Comprehensive selection
4. Continual selection
   Simple continual response (neutron bombardment, as in the atomic bomb)
   Compound systematic and complex continual response (as in digestion or coagulation of the blood)
   Solvate selection
   Superhigh-speed and compound-selection continual response (as in mammary gland activity or the photochemical reaction of the eyeball)
5. Complex unifying selection involving groups
   A. Correlatively systematic selection (as in cell-nucleus chromosome genes)
   B. Compound correlative continual selection (as in the brain tissue structure)
   C. Correlatively comprehensive and systematic transpositional selection (as in the human body)
   D. Self-determining, self-conscious selection (as in human consciousness)
   E. Phylogenesis

Classification of selection

There is a difference between selection that takes place unconsciously and selection that takes place consciously. The natural selection that occurs in the world of nature has traditionally been referred to in Japan as a *sifting and sorting out* that occurs in nature blindly and disorderedly. The closer one studies the process, the clearer it becomes that it is neither blind nor disorderly. There are three kinds of selective principles in the inorganic world, the organic world, and the world of life, each with its own functional mode.

Even in the case of the laws of selection ruling the inorganic world, we cannot speak of a blind and disordered process. As the Russian chemist Dmitri Mendeleev pointed out, a strict arrangement exists that takes the form of a kind of "periodic table of the elements." The non-compound elements are divided into nine groups (with the six types of non-compound atoms forming a group of its own). Compound elements are separated into eight groups with the fourth at the center. Groups one to three are marked as *positive* and are compounded with a plus (+) electrical charge; groups five to seven are compounded with a *negative* or minus (−) charge. Group eight exhibits a particular compound formation of its own, as in the case of the metals.

What is more, the law discovered by Italian scientist Amedeo Avogadro manifests an amazing uniformity in the spatial capacity of molecules.

In the world of organic chemistry, the selection principle explains a still stranger effect: In the inorganic world, solids, liquids, and gases separate according to differences in temperature, and yet in the organic world, carbon and nitrogen compounds show that temperature makes almost no difference in solids, liquids, and gases, thus laying the necessary groundwork for the appearance of "life" on the planet.

In this way, with the emergence of selective activity by living things we see the appearance of laws governing reproduction, regeneration, and restoration, as well as selective activity for the sake of metabolism, growth, and evolution. Nevertheless, the struggle for food leaves us with a sense of a cruel selection going on in the "survival of the fittest" where the weak fall prey to the strong. And yet, recent developments in ecology have drawn our attention to the discovery of the presence of other restrictions and arrangements.

Selection in physical chemistry

Simple selection (energy levels and molecular diameter in atomic structure, diameter of atomic molecules), configuration selection (ortho-, para-, meso-, cis-, trans-), bonding selection (nuclear and electron bonds, dual bonding), bond angles, atomic and molecular weight, atomic and molecular numbers, elementary particle numbers, isotopes, bonding proportions, light waves and ray length, directionality of very high frequencies, temperature; rejection selection (Pauli laws of atomic structure), typological selection (amino acids, proteins, nucleic acids, etc., compass points, pH ion density, cooling crystallization point, crystal axis, liquid crystal, liquid density, viscosity coefficient, melting point, boiling point, sublimation point, magnetic intensity, interface, surface tension, centripetal force, centrifugal separation, solubility, transparency, sound propagation level, conductors, conductor-seeking, colloid magnetism, particle bonding, electron bonding, molecular bonding, saturation), object selection (electric poles); complex selection (the buildup of an electron shell, cell mitochondria), associative selection (histone), catalysis, hypallage (replacement, carrier), collodification, secure orbit (inner-atomic, astronomical), circular orbit, elliptical orbit, parabola, hyperbola, circinate, latent heat, expansion, degree of condensation, cohesion, closure, opening (law of cause and effect), order, chain reaction, electron and molecular rings (nitrogen, carbon, oxygen), Krebbs ring, seasons, ellipsis, polymerization, contrast, copying, organic chemistry, wave length, acoustic harmony, ultrasonic amplitude, solid electricity, insulation, semi-conductor electricity, particular selection of osmotic force, electrical generation, optical polarization, in vivo balance, homeostasis, oxygen flow operation, enzyme activation, metabolism, catabolism, degree of freedom of transformation (Gibbs phase rule of ATP transformation with phosphorus), intracellular biology, heliotropism, negative phototropism, hydrotropism, cell-nucleus splitting, botanical blooming, aerobes, anaerobes, negative ion affiliation, positive ion affiliation, adsorption, desorption, magnetic-chemical gradation, absorption, anti-absorption, bases, acids, proteins (globulin proteins alpha, beta, gamma discriminant; bacteria identical to organisms that feed off only levorotatory proteins and exclude dextrorotatory proteins), starches (alpha and beta starch discriminants), fats, blood level, blood type, blood pressure, emulsification, vitamins, hormones, water pressure, sensation.

There are many more examples of selection that could be taken up and my insufficient research into the matter may have left out some of the important ones. I am thinking in particular of the exercise of selection in the struggle for food, as in the phenomenon of certain insects that feed off of specific plants, or of those parasites that choose only certain hosts. These examples and more could be listed, but I will stop here to avoid making matters too complicated. Suffice it to note that a solid path leading to the realm of purpose may be detected, and that even in the realm of unconscious choices, not everything is without order or limits.

Opening and closing in natural selection

Selection bundles

There is a regulating activity behind the arrangements of natural selection that enables given phenomena to continue in existence. Consider, for example, how sea water is kept above the 4° freezing temperature or how the body temperature of the higher animals and the density of pH ions is maintained. All these things are the result of a regulating activity completely dependent on combinations of nature's principles of selection. In this way, cells, the basic units of living organisms, are organized and given the potential to "exist." Now their existence may have been regulated by selection, but this does not mean they have the potential to evolve. This is where the question of opening and closing in selection enters into the picture.

Evolution by sexuality shows an opening in selection. Living organisms began evolving this way when seaweeds leaped out of the sea and onto dry land and angiosperm was born from gymnosperm. Animal instincts mark a closing in selection; developmental selection marks free will.

Closing and opening in selection

The enclosure of active energy within a certain frame is called *matter* and its opening, *light*. When atomic bonds are framed in a certain way, we have *molecules* which are units of *chemical compounds*.

Life manifests a certain durability in the midst of instability wherein all the rules of energy combine to concentrate the variability of solids, liquids, gases, electricity, magnetism, radiation, and so forth into a single point. Where this variability is absent, we have *death*. Biological instinct is manifest in this life force as that which energizes temporal selection and brings about a self-liberation. Only when selection has the energy to gather itself together in this way can *purpose* enter onto the scene as an ideal.

As for the "universe," we might think of it as a realm that has bundled up the laws that keep selection enclosed. From this perspective, the development of consciousness and the embodiment of a living conscience may be considered a liberation from the closed world of selection and the effort of life to reach out for a new world of *purpose*.

Modes of opening and closing in selection

The closure of selection gives rise to mechanization and reveals a cyclical form of energy. With the opening of selection, purpose ascends to consciousness, and through the progress of consciousness, purpose also advances. Adaptations and adjustments arise in response to differences in the form this closure takes, and life is made possible by establishing connections among the many chains of things that had been closed off by selection. When the selective closure is complete, we have a machine. The closure of "energy" within the universe as a whole is referred to as the *conservation of energy*. In the case of materialized energy we speak of the *conservation of matter*.

Varieties of simple selection

There are two systems of simple selection, the negative and the positive. On the one hand, there is a selection of temporal stasis in which an entity or organism in state A or position A stays put from one moment to the next; on the other, there is a momentary transition to state B or position B. Granting the fact that the simple selection of "+" or "−" is necessary in a decidedly mechanistic world, the actual work required is tremendous. And yet the fact remains: simple selection is going on here and now in the world of nature.

Within the atom there are protons (+) while electrons (−) circle about it on the outside. Electrodes are distinguished as plus or minus, and magnetic fields also show a plus or minus directionality. The same is true of chemical bonds, and in the case of crystallization, the directionality is decisive. When we come to the world of living things, sexual differentiation stands out above all else. Viewed in terms of electricity, it is interesting to observe the role that plus and minus play in differences between man and woman, positive and negative, male and female: intracellular nucleoplasm is "+" and extracellular protoplasm is "−".

From the standpoint of ligand field theory, the sun is plus and the earth is minus. Or again, taking the constellation Sagittarius as a galaxy in the solar system, if the surrounding area is a plus and revolves around it, then we should think of it as a moving thing with a minus form. As simple as this sounds, the organization of the cosmic mechanism through simple selection is unimaginably complex. In terms of *energy*, simple selection becomes directionality, the "+" and "–" of axial selection. *Change* involves the plus and minus of chemical combination; *development*, a choice between moving forward and standing still. In the world of *life* sexual differentiation appears; in the world of *law*, a choice between *affirmation* and *negation* comes into play. Here again, as simple as this may seem, when we combine the work that goes on at the level of energy, change, development, law, and life with simple selection, we have a world whose complexity defies imagination.

In the game of baseball, there is an *energy* in the ball that determines if it is a strike or a ball, if it is hit or missed; on the field, either first base is reached or it is not; in the *development* of the game, the process involves singles, doubles, triples, and home runs; compliance with *laws* or their infringement determine things like fair and foul, safe and out, and decide whether a run has been scored or not.

The interesting thing is how such combinations of simple decisions bring about purposive behavior that make for such a splendid game.

Laws governing selection

Some modern biologists reject the idea that there is law involved in the selection principle. For my part, I recognize a principle of selection grounded in laws present in the natural world. I recognize that things like origin of the "species" did not come about by chance but depend on laws latent in the universe.

It seems to me that the fact that all members of the insect family have six legs and all mammals have seven cervical vertebrae points to the presence of a strict arrangement in the natural world.

In my view, energy wells up within living organisms orthogenetically as a matter of methodical, systematic generation, while outside of them, a work of selection takes place both through normal and abnormal change seemingly brought about by chance.

I cannot believe events in the natural world are all products of chance. Even if natural events are blind, ultimately they manifest a finality that developed into human consciousness and hence necessitate the examination of the various points at which blindness and finality are connected.

My analysis may be insufficient, but I find the following laws latent in the natural world to govern the phenomenon of selection:

> fundamental laws of selection / the relativity of selection / selection and natural variation / mutation / selective connections / selection unification / selection accumulation / selection aggregation / selection concentration / selection differentiation / selection time-saving / selection abbreviation / selection mechanization / selection adjustment / selection balance and homeostasis / selection compensation / the transition from a principle of selection to a principle of purpose (inclusive of all principles)

In what follows, we will touch briefly on each of these laws of selection.

Fundamental laws of selection (Mendeleev's periodical table)

Each atom has its own atomic value, according to which the elements are divided into eight groups—nine if we include the non-compound atoms. When he first came up with his "periodic table," Mendeleev was excited at having "discovered an essential intelligibility in the universe." Actually, the "principles of chemical compounding" governing the elements represent "principles of adaptability" that oblige us to think in terms of a selection at work even in adapting to the inorganic world. These elements are divided into two types, plus and minus, each carrying out an adaptive selection that deserves attention.

The relativity of selection

Atomic value is the fundamental principle of selection, but it has a relative character. Hydrogen has a "plus" value towards an opposing element, but it can also change to a "minus." In combination, oxygen usually has an atomic value of two, but depending on the situation it can take the form of SO_2, SO_3, or CO_2. Even in the case of iodine we find both I_2O_4 and I_2O_5; or again both the compounds $NaIO_3$ and $NaIO_4$ exist. By examining one by one the forms taken by elements when they are compounded, we come to see their relativity.

The directionality of selection

When it comes to selection, we always think of directionality, tendency, velocity, and the like. The direction of a magnetic force is determined by *selecting* the direction of an electrical current but also points to an *inclination* towards that direction. Selection is present in the direction in which wind and electrical waves move, which greatly affects their ability to act. What is more, by maintaining their direction, light waves create the particles that make matter possible. Atomic motion combined with velocity becomes light and generates heat. Because this motion is stable in things with a circular or elliptical orbit, its direction does not change without a special reason.

Selection and natural variation

Without natural variation there would be no selection. Norman L. Bowen's research in petrology focused on the way in which differences in temperature can shape igneous rocks and bring about crystallization through cooling and subtraction. In contrast, E. A. Irving described how metamorphic rocks are produced by increasing pressure and controlling temperature. Earlier we alluded to Gibbs's metallurgical "phase rule" that proposed a "degree of freedom" in the transmutation of solids, liquids, and gases. Such variations in nature are the cornerstone of natural selection.

Selection and mutation

Mutation appears in the world of living things. Crossbreeding easily produces mutations, and reproduction through the planting of cuttage preserves the fixed form of a plant. Differences in temperature and the use of drugs enable plant mutations to be used in thremmatology. It would seem that some plants allow for mutations and others do not. The dahlia is an example of a plant that takes the striking form of a mutation. The addition of human protection can also effect mutation, evidenced by variations occurring in a large number of plants and animals.

Selection connections

How do things connect? Connection belongs to the realm of selection. Positrons gather negatrons together and quanta of light combine to make elementary particles. The amassing of various sorts of these particles form atoms and molecules, and the molecules combine to create organic entities, whose compounds become the cells that make up the higher organisms.

The direction of electrical current and magnetic force represent the most marvellous form of selection connection.

Within the organic realm, selection displays profound correlations. The eyes of white-haired animals have a red luster, and growth requires

Natural Selection and Directionality

the combined efforts of adrenalin and the thyroid gland. Or again, the salivary glands have a deep connection with the hormonal excretions of other digestive organs. As saliva enters the stomach, pepsins are set in motion that direct the activation of pancreatic juices, bile, and so forth, as well as the regulation of starches, fats, and proteins in the liver. As bile is excreted, the pancreatic enzyme steapsin begins to act. Such astonishing interconnections exist within the bodies of living things.

Selection unification

A latent unity is to be seen in the various forms of selection in the universe. Nebulae have the capacity to generate electricity through whose magnetism gravitation takes place, planets revolve about fixed stars, and life forms come to birth on the planets. There is a remarkable supervisory force latent in the birth of these living organisms, such that even the small amount of waste produced within an egg becomes a useful resource in the evolution of life. Muscles are organically connected to hundreds of bone fragments, and there is a surprising connection between the two hemispheres of the brain such that even if the left hemisphere is lost, the right hemisphere can manage speech on its own. The unity of selection refutes the idea that nature does its sifting blindly.

Selection accumulation

Charles Darwin drew attention to the imperfection of the eyeball. There is no doubt that coordinating the accumulation of choices appearing in each part of the eye and extending it to the retina is an immense enterprise. Take the refractive index of the cornea and of the crystalline lens, the coordination of the distribution of derivations in the neural network and chromatic optics of the retina, the mode of optical diaphragming, adjustment of the curvature of the lens—just adding up the selections of these things is a considerable chore.

But a still more complex accumulation of selections can be seen in the order in which an organism is generated from an egg. Tens of thousands of choices pile up in the birth of higher animals like mammals. This accumulation is not something a theory of blind sifting could support.

Selection aggregation

The work of enzymes also entails compounding, as when auxiliary enzymes are attached to enzyme activators. Turning to physiological relationships, the respiratory system serves as the mechanism for speech and the urinary system is used for reproduction. The nervous system that controls sensitivity to touch, heat, cold, pain, and pressure aggregate in the skin. Numerous such examples can be seen at the physiological level.

Selection concentration and synthesis

In addition to accumulation and aggregation, concentration is manifest in the work of selection. The structure of crystallization in the inorganic world, optical and acoustical concentration and harmony, and in the organic realm a variety of physiological systems, beginning with the skeleton and muscles, display a concentration and synthesis of selections.

Selection specialization

Cells divide and differentiate into exodermis, endodermis, and mesodermis. Each dermis or layer of skin is differentiated orthogenetically for each of the various physiological systems.

Not surprisingly, this differentiation is also to be found in every kind of elementary particle in the inorganic world. From an astronomical viewpoint, cosmic dust is differentiated into fixed stars, planets, and satellites, and the process may be thought of as structured in a manner suitable to the generation of life.

Selection time-saving (work flow)

When we observe in detail methodical (systematic) generation through selection, we are struck by the temporal reduction that occurs in the work of enzymes. Consider the fact that in the hatching of a chicken from an egg, the billions of years involved in the progression from initial cosmic dust to the evolution of life is shortened to a mere twenty days, and in the case of human beings the time of gestation required from egg to mammal takes place within ten months.

Selection abbreviation

In the systematic generation (phylogenesis) of a chicken from an egg, we note an abbreviation of several steps within the egg. Considering that a process that took five hundred million years of continual evolution from the Cambrian period is completed in twenty days, we understand how many steps of selection have had to be skipped over to abbreviate the task. This explains the rush to theories of creation from an original seed stock. Following the way quantum theory and wave theory are harmonized within wave mechanics, we may consider evolution to be a continual wave and creation a quantum node.

Selection mechanization

Not only in the accumulation, concentration, and synthesis of selection, but also in its differentiation and work flow, the need arises for mechanization. Beginning with periodic variations in luminosity in variable stars, periodic cycles in geothermal energy, and the horizontal and vertical convection, circulation, and so forth that occur in air, we cannot fail to notice the working of a comprehensive mechanism coordinating all localized selections carried out in the conscious and unconscious activity of living organisms, including the mechanical compounding and metabolism involved in the generation of organic systems.

Selection adjustment

Adjustment is necessary for selection to fulfil its goal mechanically. A large number of adjustments come into play in maintaining circulation of the blood by the heart. Adjustments are made in the concentration of hydrogen ions, body temperature, and the osmotic pressure of water, salt, and sugar in the blood. In addition, adjustments of the various internal organs as well as the sensory system are all tuned to measure moderations in selection.

Selection balance and homeostasis

High and low pressure need to work together in order to achieve balance in the earth's atmosphere. Moreover, as Avogadro has pointed out, compound molecules with the same temperature and pressure maintain an identical volume. This is an amazing balancing act. We may call it a standardization of the law of natural selection. Walter B. Cannon has drawn our attention to the achievement of homeostasis within the body. Balance is related to space and homeostasis to time. Of all natural selections, balance and homeostasis originate in the most extraordinary and sublime of selective activities.

Selection compensation

A remarkable fact of the organic world is the way in which compensation is made for an injured part of a living organism. When an injury occurs

Cosmic Purpose

to one part, a neighboring part compensates for it, or, as happens with blood, acts on behalf of the whole to effect the compensation.

To counter the "death" of one individual, the work of reproduction is present to plan the restoration of the totality of the individual body.

According to Fred Hoyle, in the inorganic world, the creation of matter through photons continues forever, and matter recovers even with the destruction of the atomic nucleus.

From a principle of selection to a principle of purpose

The presence of adjustment, balance, homeostasis, and the like in the work of selection is an astonishing phenomenon of the natural world. Still more astonishing is the advance in the organic world to a selection principle involving three or more options. The techniques used by bees and ants to communicate a destination to their friends fully merit consideration as an advance from a selective principle to a purposive one, while the exercise of consciousness in higher animals surely demonstrates the transition of selection to a high level of purpose.

Oparin's materialism

The Russian biochemist Aleksandr Oparin attempted to affirm finality from the standpoint of dialectical materialism. Rejecting metaphysical explanations of any kind, he believed he had found a way to confirm, through the laws of physics and chemistry, an evolution in the world of living organisms from the inorganic to the organic that transcended biochemistry. He considered the order of this evolution to be a matter of systematic generation that came about through spatial harmony and temporal tuning.[6]

Oparin thought of the world of life that made its appearance in a lawful and purposive world of living organisms as a kind of music that developed over time,[7] from the first stage of carbohydrates, to the second stage of amino acid compounds, to the third stage of the sphere of enzymatic proteins. Oparin's thoroughgoing affirmation of finality dismissed Aleksandrov's claim referred to earlier, that those who believe in finality are "the enemies of science." He was one of the rare few with the cour-

---

6. See A. I. Oparin, *The Origin of Life* (New York: Dover, 1952).

7. See the lecture of Prof. Oparin's transcribed in the proceedings of the Japan Society for Life Studies, 『日本生命学会』 November, 1952: 27.

age to try to demonstrate through his research on enzymatic proteins a materialistic form of finality. He represents an interesting chapter in the development of modern biology.

Those engaged in enzyme research, like the German chemist H. Emil Fischer, thought in terms of "locks and keys," but Oparin took the bold step of introducing "teleology." In so doing, *atheistic* teleology was transformed into an *immanent* teleology, leaving no option for a *pantheistic* teleology. We cannot fail to note the fact that Oparin's approach surmounted the logical contradiction in the claim that a purposive world of enzymatic proteins could come to birth from a chaos lacking in purpose; it sought to construct within Russian dialectical materialism a view of the cosmos with purpose.[8] We see the true face of his atheistic teleology when he writes, "When we actually investigate primal matter, mechanical principals are refuted. Nothing at all like a mechanical principle is to be found in primal matter."[9]

Statistical mysteries

Selection surfacing in the statistics of inorganic entities

Statistical research is rather like chewing sand, and yet one feels, as with higher algebra, that there is something interesting to be learned. Like tracing a coefficient back to its root, fishing through statistics and hitting on things in the figures like regularities, ratios, probabilities, constants, and a priori homeostasis, gives one the sense of some mystical control deep within the world of numbers.

Statistical research in both the inorganic and organic realms draws attention to the way in which "chance" is ruled by "field," "selection," "process," and "law." One senses within the recesses of so-called chance events a certain arrangement with the surrounding field and certain limits to "change" that rule out "unrestricted chance."

When Sommerfeld, to whom we owe the introduction of "selection rules," was conducting statistical research on the way electrons fly off their normal orbit, he discovered selection at work in energy levels. The Danish nuclear physicist Niels Bohr startled the academic world by pointing

---

8. Oparin オパリン, 『生命の起原』 [The origin of life] (Tokyo: Iwasaki Gakujutsu Shuppansha, 1968), 69, 73, 74, 75. [The English translation cited in note 4 is not coincident with the Japanese; in this and the following note, the latter is therefore cited.]

9. *Ibid.*, 76.

out that all the stable places in an electron's orbit appear in energy levels that are integral multiples of the electron's diameter. Sommerfeld took the study of the electron orbit further by introducing finality into metallic crystals.

Avogadro's mysterious research into the number of molecules

In 1811 Amedeo Avogadro (1776–1856) advanced the hypothesis that "in the same temperature and under the same atmospheric pressure, identical volumes of all gases contain the same number of molecules." Tanaka Minoru states in his *History of Modern Chemistry* that over a period of some one hundred and forty years the hypothesis gradually came to be treated as actually proved.

In 1874, sixty-three years after Avogadro had set forth his hypothesis, George J. Stoney's calculations led him to the theory of an elementary charge, that is, the electrical character of a monovalent ion calculated by dividing the chemical equivalent of one Faraday by the Avogadro number. In 1891 this elementary charge was given the name "electron." In France, it was Jean Perrin who measured the number of molecules. In 1909 Perrin, by analyzing the movements of micella in emulsion particles of olibanum, looked for the Avogadro number from the submerged equilibrium of the particles. Confirming the value by other methods, Avogadro's hypothesis was proved to be a matter of objective fact.[10]

I find interesting the way in which the essence of selectivity governs the universe such that the total number of molecules in bodies of identical volume, temperature, and atmospheric pressure, regardless of what atoms they are made of, should be exactly the same.

This is but one of the statistical mysteries of inorganic entities.

Mystery in the measurement of amino acids

Vladimir Vernadsky, the professor of Moscow University who founded the field of geochemistry, was a chemist and a theist. Expelled from the university under the atheistic Stalin, he taught at the University of Paris during the Second World War. Operating from the principle that colloidal matter in the earth's crust easily converts to living matter, he judged that the gross volume of colloidal matter on the earth crust is constant, and therefore that the gross volume of living matter is also constant. That

---

10. Tanaka Minoru 田中 実, 『近代化学史: 化学理論の形成』 [A history of modern chemistry: The formation of chemical theory] (Tokyo: Chūkyō Shuppan, 1954), 305.

would lead us to conclude, for instance, that the total volume of locusts born in a desert region approaches that of colloidal matter in the region.

Research involving this kind of quantitative statistical estimate is interesting, but recently I have become intrigued by advances in biochemistry, where, further research on amino acids has led scientists studying proteins to begin speaking of "finality." It is interesting to recall that Aleksandrov dismissed those who subscribe to finality as "enemies of science,"[11] but Oparin, while believing in dialectical materialism, accepted finality from the perspective of biochemistry. Studying the sequence, structure, and orderliness of amino acids, he said he could not avoid acknowledging finality.

Scientists conducting statistical research on amino acids have noted this finality. Living organisms make use of chemical energy produced in the course of oxidation reduction. However, because heat is dispersed in ordinary combustion, rendering it unfit for activities associated with life, a direction conforming to a special design is taken. Namely, several enzymes with an "orderly electric potential difference" are assembled and a stored energy is contrived that allows for at least some of the chemical energy to be used.[12]

Along with such statistical studies in quantum chemistry, when population statistics of living organisms are investigated, something startling is discovered. On a recent trip to Japan the German biochemists A. F. J. Butenandt and Josef Leinen each made presentations concerning biochemical research on genes and protamines. During their talks they recounted a statistical mystery, observing that a finality has been contrived at the bottom of the microscopic universe, by means of which a marvelous, purposive design is present in the world of genes that supersedes the mechanism of the macroscopic world governing the compatibility of males and females. Looking at population statistics after the Great War, the percentage of male children born was comparatively higher than that of female children, a phenomenon that continued for several years thereafter until a general parity had been reached and the birth of male children subsided. I find this astonishing.

And there is more. Organisms with little strength for combat have an extraordinarily strong reproductive potential, which leads us to think

---

11. See note 1.

12. Ōki Kōsuke 大木幸介, 『生物量子化学』 [Quantum biochemistry] (Tokyo: Kyōritsu Shuppan, 1950), 12.

that even if they lack the power to defend themselves, their species can be preserved.

If ungulates and other animals that cannot protect themselves produce male and female offspring in largely equal quantities, their ability to reproduce at a very high pace lessens the difficulty of preserving the race from destruction by predators. Such statistics concerning the reproduction of living organisms seem to me to point to the mystery of selection.

The Development of a Telology of Selection

Laws governing the selection of an electron orbital

There are seven orbitals or "shell" areas for the ninety-two chemical elements. Their names are assigned from the inside out: (1) K, (2) L, (3) M, (4) N, (5) O, (6) P, and (7) Q. These shell areas are further subdivided into orbits referred to as L. Things are arranged in such a way that the number of L orbits changes depending on whether the number of electrons in the atomic structure is large or small. Thus rare gases like radon, whose atomic number is eighty-six, have a series of seven sub-orbits at N, while another rare gas xenon has only five.

Even so, considering the ingenious way the atom is structured, we have to stand in awe at the "laws of selection" governing electron orbits. We usually think of electrons as being free to transpose or transition each of their orbits, but when they actually transpose or change orbits, invariably there are places that must be avoided. Counting from the innermost nucleus, the shell areas mentioned above are named as (1) S (sharp), (2) P (principal), (3) D (diffuse), (4) F (fundamental), (5) G, and (6) H.

When an electron shifts from one orbit to another, it always shifts to another shell area and must avoid things with the same order number as the sub-orbit of the previous shell area. It must also move away from both sides neighboring the order of the previous shell. In the case of electrons, we cannot think, as the Japanese proverb has it, that "shame goes out the window when you have crossed the border." In the case of the orbit of a neighboring shell, there is no freedom to jump in without concern for the order. Wandering into L+1 and L-1 is always strictly prohibited. In atomic physics this is called the "selection rule" of electron orbits.

To take the example of alkaline earth electrons, the only movement for S is to the P order of each shell, for things in the P order to the S and D orders, and for electrons of the D order to the P or F orders.

Natural Selection and Directionality

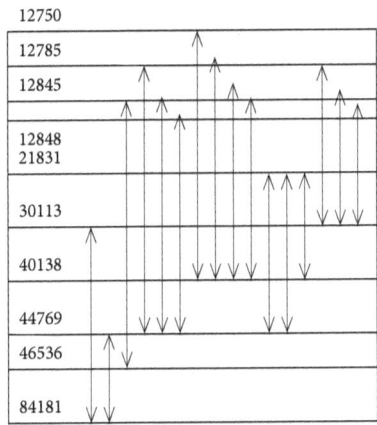

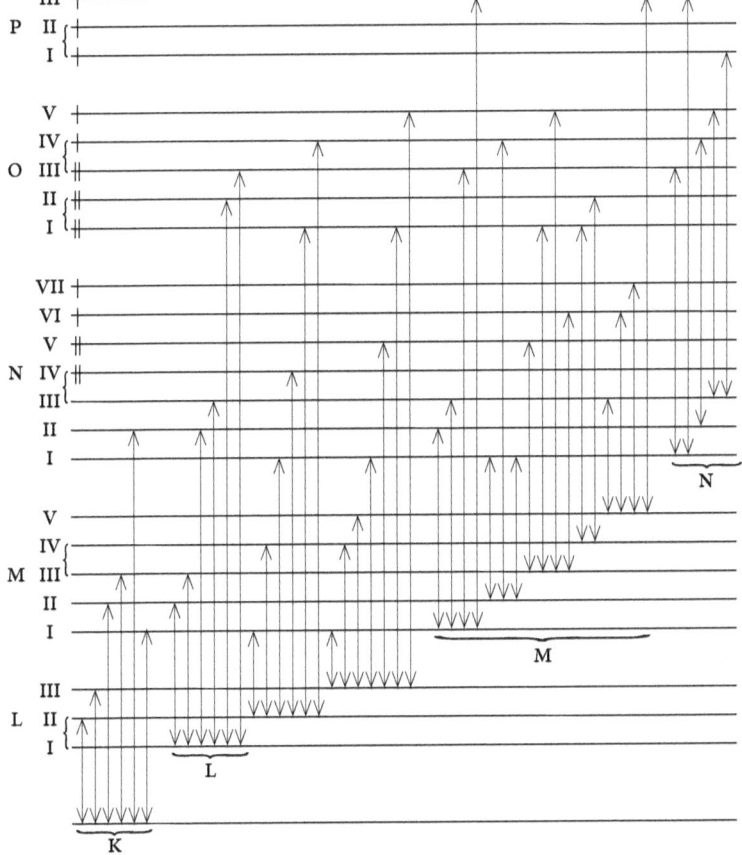

Selection in the energy levels of electrons

Cosmic Purpose

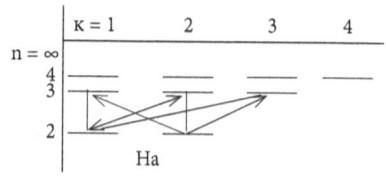

Detailed structure of hydrogen atom energy levels

Of course, a transposition resulting from the collision of electrons represents an extraordinary case in which the rules do not apply. But the general way in which the rule is upheld is something of a marvel.[13]

The selection principle in the transition of electron levels

Following the Rydberg-Ritz combination principle, in the diagram above, *Ha* should have six lines, but in fact it is composed of only three. This is due to the "selection rule" limiting the transition of electrons or change in quantum number. In the present case, K is permitted a transition of only one level. Thus, the three lines 2S–3S, 2P–3P, 2S–2D appear in the spectrum, but there is no transition because the difference of K is 0 or 2. (In the figure above, the dotted lines indicate what is not possible.)

In theory, however, this selection principle is not so strict as to exclude the possibility that lines forbidden by the principle can be drawn in cases like the atomic spectrum of a large atomic number.[14]

The 2573Å of Mercury (Hg) and the 6573Å of Calcium (Ca) both represent a transition from $3P_1$ to $1S_0$ which is something that cannot occur in hydrogen because of its low atomic number.

Selection in the absorption of special wavelengths in molecular oscillations

In studying the structure of ammonia, we see that it consists of three molecules of hydrogen and one of nitrogen. Furthermore, when the three hydrogen atoms are set on a plane, they make an equilateral triangle, and with the addition of the nitrogen atom, a tetrahedron. With this plane as a base, nitrogen is positioned at the apex of the tetrahedron. The interval between each of the three hydrogen atoms each is 1,628 angstroms (a unit angstrom being .0000000001 meter). The distance between the

---

13. Linus Pauling and Samuel Goudsmit, *The Structure of Line Spectra* (New York: McGraw-Hill, 1930), 49.

14. See Azumi Hiroshi 安積宏, 『量子化学』 [Quantum chemistry] (Tokyo: Baifūkan, 1951), 28–9.

nitrogen and hydrogen atoms is 1,014 angstroms. Accordingly, the distance of nitrogen from the base is about 614 angstroms shorter than that between each of the hydrogen atoms. This nitrogen atom covers the base and moves back and forth methodically in a plus and minus direction. In physical terms, it oscillates.

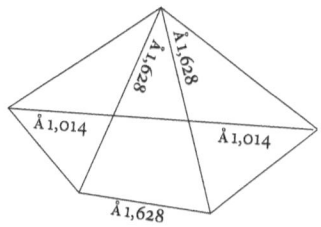

Taking this oscillation and applying it to a quartz clock gives us an atomic clock. Oscillations of 23,870.16 megacycles (billionths of a second) per second are absorbed.

We should note here that ammonia gas absorbs wave frequencies oscillating at this rate. What is interesting is that this particular absorption is used to make accurate adjustments in the electric oscillations of the quartz. By constructing a thin quartz plate and adding pressure to it, electricity is produced on the surface. In contrast, the plate expands and contracts if voltage is added to it with a silicon film. That is, oscillation is initiated. This is referred to as the "piezo effect." In crystallography this oscillation was used in the creation of the quartz clock, which is second in accuracy only to the atomic clock.

To begin with, the waves that create oscillations in a quartz clock are passed through a tube with ammonia gas. When this happens, ammonia absorbs only oscillations of 23,870.16 megacycles a second. The absorbed oscillations are then passed through a quartz clock, which adjusts the clock's oscillations to a strictly defined oscillation. This by itself assures a variance of about one second every six years. It is interesting to note here that the use of such a clock allows for the velocity of light to be calculated.

What I mean to draw attention to here is the presence of a strict, physical selection in the universe and the fact that the combination of various selections gives rise to a purposive mechanism. Clearly machines are compositions of selectivity in which, let us not forget, a purposive world is manifest.

The development of purposive teleology

During my investigation of the selection principle, I noticed that what Albert Szent-Györgyi referred to as "muscle research" does not oblige us to look for a purposive teleology outside of the selection principle. Moreover, questions about muscles or life cannot be solved with Hegelian dialectics alone as the dialectical materialists claim. Even if we consider

Cosmic Purpose

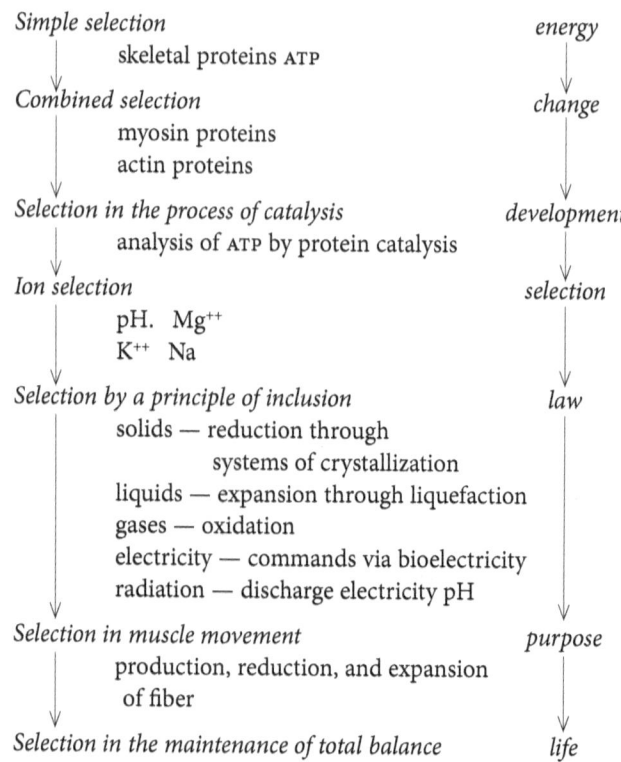

only contraction and relaxation, muscular theory has never progressed in a dialectical fashion.

The selection principle itself displays a purpose regulated from the simple to the complex as selections combine and compound, putting up resistance and sustaining balance as the need arises—all in the service of a methodical and flawless advance. There is no real need to seek purpose outside of the mechanism's own modality. The machine moves with purpose because the factors involved in its mechanical operations belong to a closed selectivity with its own *internal purpose*. (The diagram above details the theory of the internal environment's adaptability.)

In the case of muscles, all matter involved in the systematization of a living organism needs energy to work. There is no reason to look for energy external to the organism itself. Nutrition may come from without, but once it is digested, it needs to function as internal energy.[15]

---

15. See A. Szent-Györgyi, *Nature of Life: A Study on Muscle* (New York: Academic Press, 1948). The contents of the adjoining table is explained on pages 32–8.

Standards for selection

As should be clear from earlier examples, standards enter the world of nature through selection. That said, to what end do such surprising standards arise?

When the manufacturing of parts is moved out of a large factory into a subcontracting facility, the standards move with it. If they were not, the parts could not be assembled. So, too, we have to suppose that standards are determined in the universe for the sake of what is constructed by bringing the parts together. Atoms are assembled from the concentration of the six elementary particles in chemical elements. Further, atoms are concentrated for the sake of assembling molecules, and molecules concentrated in assembling cells. For metabolism to take place, there must be something like the standardization of parts that Ford Motors uses in its automobiles to insure that by changing parts an older car can run like new.

Put this way, the system of standards we find in the natural world is remarkable. But just how could the choice of standards have been decided on? We can only think that there is a hidden wisdom within the recesses of the universe, arranging things behind the scenes. Wisdom commits itself to standards, and it is precisely when the universe is unified through standards that the movements and developments of the natural world aim at evolution.

H. G. Wells once wrote that by throwing around the letters of the alphabet at random, sooner or later one would come up with the works of Shakespeare. But has he really taken into account the *selection* of alphabetical standards and the *aim* of reproducing Shakespeare? Without standards of alphabetization, Shakespeare's works would never have been written, no matter how long things were left to "chance." And even if the texts *were* so reproduced, they would be illegible without alphabetical standards. Still, Wells considered standards unnecessary; and he further believed that if great literature is possible, then literary language should also be also possible even without written letters.

2 Standardization in Natural Selection

Natural selection in mathematical integers

The ancient Egyptian philosopher of mathematics Ptolemy tried to ground philosophical principles in numbers. Strong hints of his influence are present in Plato and Aristotle. I would like to recall those traces and then take the further step of obliterating all superstition surrounding numbers so that, grounded in the image of the universe we have received from modern physics, we may discover a new standard for integral ratios from the viewpoint of the principle of natural selection.

I do not wish to make the demanding ideas of the philosophy of mathematics my starting point. Rather, I would like to begin by taking up some of the integers, proportions, and series that atomic physics, astrophysics, biophysics, and so forth have given us and try to learn what they mean.

The emergence of quantum integers

In the year 1900, as the new century was getting underway, Max Planck, a professor at the University of Berlin who was engaged in research on heat radiation, discovered the constant 10^{-27}, ringing in a new dawn in modern physics. He had found that heat radiation did not depend on chance frequencies but clustered into multiples of integers forming fixed bundles. The news came as a heavy blow to those seeking to build their views of the universe only on arguments from chance.

Planck's radiation integers represented the truly startling discovery that when h is disclosed, the ratio of nh applies throughout all substances in the universe. Thus it became clear that electrons and quanta, as well as molecules and all chemical compounds, are nh multiples. The standard character of this integral ratio offers a standpoint from which to study the selection principle in nature and from there pursue other forms of standardization.

INTEGRAL STANDARDS IN NATURAL PHENOMENA AND THE PRINCIPLE OF NATURAL SELECTION

The following are examples demonstrating the presence of an integral standard in inanimate matter:

1. Dalton's theorem of multiple proportions
2. Gay-Lussac's theorem on multiple proportions in gases
3. Integral ratios in the expansion of gases
4. Avogadro's theorem on the volume of molecules, etc.
5. Prout's theory of the atomic weight of elements as an integer multiple of hydrogen
6. Faraday's constants
7. Planck's thermodynamic quantum constants and the principle of integer multiples
8. Integers appearing in the atomic realm
9. Integral ratios in atomic spectrum light rays appearing in Moseley's law
10. Integral ratios appearing in celestial bodies
11. Integral ratios appearing in music

I should like to describe each of these briefly.

Dalton's theorem of multiple proportions

Atomic theories were first formulated in modern times through the experiments of John Dalton (1776–1844). It did not occur to him to ground the discovery of his formulations in a theory of materialism. He was a fervent religious believer, born into a religious family and raised in simple poverty. His father worked at the wool mills and his mother at a stationery store. Both belonged to the Quakers in England and as such took the opposition to war seriously. The father of modern atomic theory grew up in this family, remained religious his whole life, and for a time ran a religious school. Even his clothes displayed the religious values of the Quakers. He never married but devoted the seventy-eight years of his life to scientific experiments.

In describing Dalton's theorem of multiple proportions, Jöns Jacob Berzelius wrote: "One may think to ignore Dalton's atomic theories, but his theorem of multiple proportions is truly mysterious." Berzelius had his reasons for saying this. It had previously been thought that interrelations among the atoms is completely lacking in order and control, and that they are ruled by chance alone, but Berzelius was aware that Dalton's experiments showed the world for the first time a systematic order among the atoms that arises in proportions that can be demonstrated through simple ratios of integers. Clearly this is not something chaotic and blind but is governed by unified and continuous laws. We recall again Mendeleev's subsequent discovery of the periodic table. Mendeleev said that when he came upon his laws, his belief in the fundamental veracity of the universe increased. Dalton's existential convictions must also have been reinforced by his discovery.

When Dalton was first engaged in analyzing ethylene (C_2H_4), he realized that methane (CH_4), in relation to the fixed weight of carbon, contained twice the amount of hydrogen found in ethylene.

This surprisingly simple relationship caught his attention. When he conducted the same experiment on chemical compounds of carbon and oxygen, he recognized that the same relationship of simple proportionality among atoms held there as well. Unlike Gay-Lussac, however, he did not experiment on proportions of volume. Unfortunately, to his dying day he refused to subscribe to such proportionality. In fact, precise experiments on the volume of gases were not possible at the time.

Dalton also experimented on nitrogen oxides. The results of his tests on gases such as nitrous oxide (N_2O), nitric oxide (NO), dinitrogen trioxide (N_2O_3), nitrogen peroxide (NO_2), and dinitrogen pentoxide (N_2O_5) confirmed that the volume of oxygen in relation to a fixed volume of nitrogen shows a simple integral proportion of 1:2:3:4:5.

By around 1802 Dalton had already embraced this theory, but he did not write it up until 1807 to 1808. In his *New System of Chemical Philosophy* he has this to say concerning the notion of the atom:

1. All elements are built up from uniform atoms possessing a fixed weight particular to that element.
2. Chemical compounds are generated when atoms of different elements bond in an extremely simple proportionality.

The results of these bondings are called molecules.

He also went so far as to publish his calculations on weight proportions among atoms. The results were later disproved in experiments carried on by numerous scientists from various countries, but the fact remains that at the time no one in the academic world was offering such theories as bold as his.

Gay-Lussac's theorem of integral proportions in gases

Among the contemporaries Dalton met with was a French scientist who was conducting experiments similar to his own, Joseph Louis Gay-Lussac (1778–1850). A professor at the Sorbonne in Paris and member of the upper echelons of chemical science, Gay-Lussac's contributions to modern chemistry include the discovery of elements like boron, potassium, and iodine.

Gay-Lussac found a strict proportion between gas volumes when they combine into a compound. He discovered that when gases A and B interact to form gas C, a certain proportion obtains among the three gases in terms of volume. Consider, for example, what happens when hydrogen and nitrogen combine to form ammonia: when a reaction is set up between three parts of hydrogen and one part of nitrogen, the proportion between them results in two parts of ammonia. Or again, when hydrogen and oxygen combine to give water vapor, the addition of one part oxygen to two parts hydrogen, two parts water vapor are produced.

This was Gay-Lussac's[1] great contribution to science.

The general rule of integral proportions in the expansion of gases

Gay-Lussac carried out additional research on the relationship between the volume and temperature of gases under fixed pressures, which led him to a general rule of integral proportions in the expansion of gases.

In 1802 he observed that when the temperature of a gas is raised by 1°, its volume expands by 1/273 of its volume at 0°. In other words, taking V as the volume at fixed pressure and t as the temperature, we have:

$$V = V_0(1 + \frac{1}{273} t)$$

The discovery of this remarkable law reveals to us an orderly arrangement concerning gases. Avogadro went a step further to disclose the orderly arrangement of molecules in the universe.

Avogadro law of molecular and other volumes

There have been great chemists like Berzelius who sensed the mystery of the universe in the simple integral ratios that obtain among atoms. But when we are informed that as more or less differently shaped atoms orga-

---

[1]. Since the name "Gay" was so common, the name of his village, "Lussac," was added by his father, a judge, to save his son from being confused with anyone else.

nize molecules and occupy a fixed space, the total number of molecules as well as their total volume remains the same, we feel as if we had been whisked off to a magical world.

In 1811 the Italian nobleman Count Amedeo Avogadro (1776–1856), a professor of physics in Vercelli, discovered the principle that states: "In gases of the same pressure and temperature and of identical volume, the same number of molecules is present."

This means that because the weight of a chemical compound is the sum total of its parts, compounds can be heavier or lighter, depending on their type. But when it comes to volume, type makes no difference whatsoever: it is always fixed. Thus, a fixed volume of hydrogen will yield the same volume of water vapor; and even if this water (H_2) combines with ethylene (C_2H_4) to produce alcohol (C_2H_6O), there is not the slightest increase or decrease in volume. To take this a step further, if one part of ethylene is bonded to alcohol, the resulting butyl alcohol ($C_4H_{10}O$) retains the original volume.

According to the research of Pauling and others, atoms are each a different length in diameter. But when they are brought together to form molecules, whether large or small, they all line up perfectly with one another. This is hardly something common sense would have predicted, but in the world of nature, miraculous things like this occur as a matter of course.

It is too much to suppose that these orderly arrangements are the result of chance. The question of chance aside, human intelligence takes it for granted that when large atoms are gathered together they should constitute a larger volume than small atoms. But in the molecular realm, a certain kind of selection is at work to counterbalance the large with the small and make everything equal. What are we to make of this?

The "mole" standard in the natural world

This marvelous phenomenon is the basis for the term *mole*. The mole, or gram-molecule, is a unit for measuring in grams the amount of a molecule or chemical compound. Accordingly, the number of molecules in one mole is always equal, independent of what matter we are dealing with. It is also equivalent to the number of atoms in an "atomic gram."

In OCL slope pressure, oxygen produces a gas of two-atom molecules with a molecular weight of 32. The volume of 32 grams of oxygen under one unit of atmospheric pressure at 0° temperature is the standard for other gases. The actual measurement is $22.4 \times 10^3 \text{CM}^3$. Following the

Avogadro principle, this means that all 1 gram molecules under 1 unit of atmospheric pressure at 0° temperature must have a volume of 22.4 liters.[2]

Prout's hypothesis

Following the discovery of the neon isotope and improvements in the equipment used for spectroscopic analysis, it became clear that all elementary particles possess isotopes. It further became clear that if we take the hydrogen atom as always equal to 1, the mass of each of the atoms of these isotopes is always an integral multiple of the hydrogen atom. In this way the atomic number came to identify the chemical mass of a given atom, displayed as a superscript to the left of the chemical symbol and referring directly to a particular multiple of hydrogen.

This hypothesis of particles as integral multiples of hydrogen was advanced by William Prout in 1815, following close on the heels of Dalton's atomic theory. Nevertheless, given the imperfect measurements of the time and insufficient determination to further research the matter, the fact that mass could be expressed in integral numbers did not make the slightest impact, and as a result Prout's hypothesis was not employed. Through the efforts of scientists like F. W. Aston in England, the study of atomic isotopes, and with it, the adoption of hydrogen as a standard, returned to theories of atomic structure and became basic knowledge across the world. Of course, as Aston and others learned through their research, not all atoms are integral multiples of hydrogen. At the same time, research on other elementary particles figuring in the structure of the atom led to the addition of a few integral functions "approximating multiple integers."

Looked at in terms of theories of the evolution of the atom, the conclusion that nearly one hundred other elementary particles evolved with hydrogen as a base represented a rich harvest of new knowledge.

Faraday's constant

In 1833 Michael Faraday discovered that when electricity disperses, the quantity of particles that precipitate to the poles is proportionate to the

---

2. See Kikuchi Masashi 菊池正士 et al., 『原子核物理学』 [Atomic nuclear physics] (Tokyo: Kawade Shobō, 1940), 9; Tanaka Minoru 田中 実,『近代化学史: 化学理論の形成』 [A history of modern chemistry: The formation of chemical theory] (Tokyo: Chūkyō Shuppan, 1954).

quantity of electricity (current × time) flowing at the time of the dispersion, and that for the precipitation of a unit of chemical weight (atomic mass divided by valence) to occur, its value has to remain fixed, independently of what the element is.

This is normally referred to as Faraday's constant: 96.94 C (coulomb). Suppose for a moment we have a univalent particle whose chemical equivalent is equal to its atomic weight and whose number of atoms is equal to the number of molecules in a mole (N). It follows that if a single atom has a certain elementary electrical charge (l_o), then $N \times l_o = 96{,}494$ C. In general with an atomic valence of N, the elementary electrical charge is the unit valence of an atom multiplied by N.

Since Faraday's time, it has become clear that the power of atoms to combine is a function of free electrons outside the nucleus, with the result that his law does not amount to much. Nevertheless, we stand in awe at the strict nature of integral ratios in the universe.

Moseley and integral ratios in light rays

The young physicist Henry Moseley who lost his life at Gallipoli during World War I offered the atomic world a great discovery in spectrochemistry by arranging frequencies in association with a physical formula that measures electrical frequencies in the orbit K:

$$V_{mo} = KR(Z - \sigma)^2$$

What a wonderful world of uniform order that makes it possible to gather all atoms into a formula covering the frequencies of given light waves! We are still further surprised to learn that those atoms are all surrounded with electrons with seven-tiered orbitals, which, because they have a length that is an integral multiple of the rotations of the electron, are arranged at a certain distance from the atom's nucleus.[3]

These seven tiers rotate by means of atoms organized under the two headings that led Niels Bohr to the following formula:

$$2N^2 \text{ and } 2(N-1)^2.$$

What this means is that there are 2 elementary particles in the first tier (K), 8 in the second (L), 8 in the third (M), 18 in the fourth (N), 18 in the fifth (O), and 32 in the sixth (P). There should be 32 in the seventh and

---

3. See Yukawa Hideki 湯川秀樹,『極微の世界』[The microscopic world] (Tokyo: Iwanami Shoten, 1942).

Standardization in Natural Selection

final tier (Q), but as the attraction from within weakens, no more than 6 can be held in check. These latter can all change location between $2N^2$ and $2(N-1)^2$. That is, supposing that N is placed at 1, 2, 3, and 4, a shift in the atomic number of uncompounded elementary particles occurs, as the following diagram illustrates:

$$
\begin{array}{llll}
2N^2 & \ldots\ldots & N=1 & \ldots 2 \\
2N^2 & \ldots\ldots & N=2 & \ldots 8 \\
2(N-1)^2 & \ldots & N=3 & \ldots 8 \\
2N^2 & \ldots\ldots & N=3 & \ldots 18 \\
2(N-1)^2 & \ldots & N=4 & \ldots 18 \\
2N^2 & \ldots\ldots & N=4 & \ldots 32
\end{array}
\qquad
\begin{array}{l}
10 \quad 18 \quad 36 \\
\\
54 \\
\\
86
\end{array}
$$

Now these 92 elements, together with the non-compound elements, are divided into 8 groups to make up Mendeleev's periodic table. Again, when he made this discovery, Mendeleev noticed a wonderful design that reinforced his belief in the truth of the universe.

The molecular "field" and design

A marvelous selective design is found in the molecular world by studying selectivity in water, oil, proteins, and other organic compounds. It is altogether startling to think that even the slightest change in a molecular equation would have a strong influence on the functioning of the human body.

Integers in the atomic world

It is common knowledge nowadays that atoms are composed of what are known as Fermi elementary particles, among them (1) protons, (2) electrons, (3) positrons, (4) neutrons, (5) mesons, and (6) neutrinos. Yet few are aware of the integral ratios that obtain among the elements that make up the atom. The discovery of the fact that the chemical processes going on among the atoms is a function of bonding among the electrons can only be called a victory for twentieth-century physics.

Electrons themselves are composed in integral proportions, but it was Bohr's contribution in 1913 that clarified the profound relationship between Mendeleev's discovery of the periodic table of elements and the various electrons that revolve in orbitals about the nucleus of the atom. This was the nearly perfect result of his spectrochemical approach to the model of the atom that had been worked out by English scientist Ernest Rutheford. Subsequently, A. E. Haas of Vienna University carried out meticulous research on electron orbitals to demonstrate a truly amazing beauty in the alignment of electrons.

The positioning of the electrons is revealed when atoms are studied under a spectroscope. Their orbits are referred to in spectroscopy as *energy levels*. They always appear at a fixed location and are never positioned by chance.

There are six elements that never compound in air with other elements: helium (2), neon (10), argon (18), crypton (36), xenon (54), and radon (86). From Bohr we have learned of the fundamental relationship between the orbitals and the number of electrons that explains just why this is so. Namely, if we take N as the orbital, then $2N^2$ and $2(N-1)^2$ and the electrons on each orbital are covered.

Table 1 depicts this graphically (and shows an interesting similarity to the order the order in which cells split).

TABLE 1. Electron clusters	K	L	M	N	O	P	Total electrons
helium · · · · · · · · · · · · ·	2	. . .	. . .	. . .	. . .	. . .	2
neon · · · · · · · · · · · · · · ·	2	8	. . .	. . .	. . .	. . .	10
argon · · · · · · · · · · · · · ·	2	8	8	. . .	. . .	. . .	18
crypton · · · · · · · · · · · ·	2	8	18	8	. . .	. . .	36
xenon · · · · · · · · · · · · · ·	2	8	18	18	8	. . .	54
radon · · · · · · · · · · · · · ·	2	8	18	32	18	8	86

K	$2 \times 1^2 = 2$	(with N as 1)
L	$2 \times 2^2 = 8$	(with N as 2)
M	$2 \times 3^2 = 18$	(with N as 3)
N	$2 \times 4^2 = 32$	(with N as 4)
O	$2 \times (4-1)^2 = 18$	(with N as 3)
P	$2 \times (4-2)^2 = 8$	(with N as 2)

As we see from the above, we may refer to the K shell or orbital as a *shell*. That is, the standard is a single energy level (or orbital) in shell K, but 3 in shell L, 5 in shell M, 7 in shell N, 5 in shell O, and 3 in shell P. These energy levels do not, however, maintain the alignment of radon (86) with respect to elements with a lower atomic number, and even when we are dealing with crypton (36), which ends at shell N, shell M fills up all of 5 orbitals and shell N ends up with 3 elements.

Standardization in Natural Selection

The alignment of electrons within the atom is by no means a matter of chance. On the contrary, it obliges us to acknowledge the presence of a wonderful law of selection. As Table 2 makes clear, electrons, with shell N at the center, disperse evenly left and right, not unlike the disposition of the planets in the solar system. We may suppose that planets with a large mass like Jupiter and Saturn are placed at the center in order to gain balance.

TABLE 2. Energy standards	K	L	M	N	O	P
cluster 1	2	2	2	2	2	2
cluster 2	. . .	2	2	2	2	
cluster 3	. . .	4	4	4	4	
cluster 4	. . .	. . .	4	4	4	
cluster 5	. . .	. . .	6	6	6	
cluster 6	. . .	. . .	. . .	6		
cluster 7	. . .	. . .	. . .	8		
	2	8	18	32	18	8
				86		

TABLE 3	K	L	M	N	O	P
radon	1	3	5	7	5	3
xenon	1	3	5	5	3	
crypton	13	5	3			

The more we study the structure of the atom, the more astonished we are at the presence of an exquisite latent design. The fifteen elements from those like the rare earth lanthanum (57) to lutetium have all been classified in group 3 of Mendeleev's periodic table. This further example of an a phenomenon necessary for preserving the equilibrium of mass in the atom amazes us yet again with the ingenious structure present in the natural phenomena of the universe.

All the elements from hydrogen (1) through uranium (92) are divided into eight groups. Their compound forms were largely determined by Mendeleev, who grew up in Siberia. Later the American Irving Langmuir tried to explain the classification three-dimensionally in terms of the eight corners of a cubic square. Although the idea turned out to be a bit impractical, it was most helpful in elucidating the phenomenon of compounding.

Cosmic Purpose

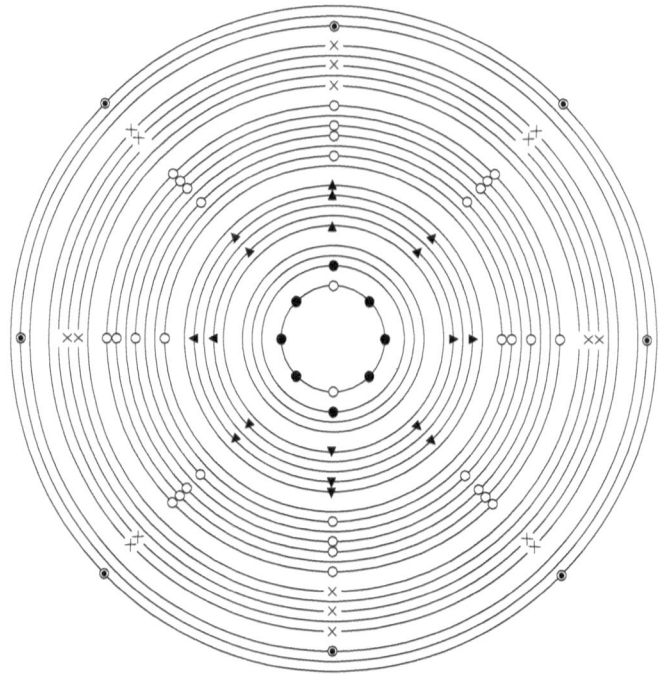

Stokes' law, named after the Cambridge mathematician George Gabriel Stokes, clarified the distribution of electrons in each shell. The graphic above depicts this in terms of Bohr's orbital circles. Today, given Sommerfeld's work, we actually have to think in terms of three-dimensional elliptical orbits, but it is simpler to imagine a series of two-dimensional concentric circles to show the complete agreement between Stokes law and Langmuir's approach.

The first group, the outermost circle, shows elements with a single atom: hydrogen (H), lithium (Li), natrium (Na), potassium (K), copper (Cu), ruthenium (Ru), silver (47), cesium (Cs), gold (79), and astatine (85). Up until about twenty years ago, astatine was called virginium, and francium, element 87, was formerly known as alabamine. Element 43, formerly masurium, is now called technetium after its discovery in the laboratory. Element 41, niobium, whose resistance to high temperatures has made it useful for jet engine turbines, was called colombium. Arranged according to Bohr's method, all of this is in line with the table on the following page.

Standardization in Natural Selection

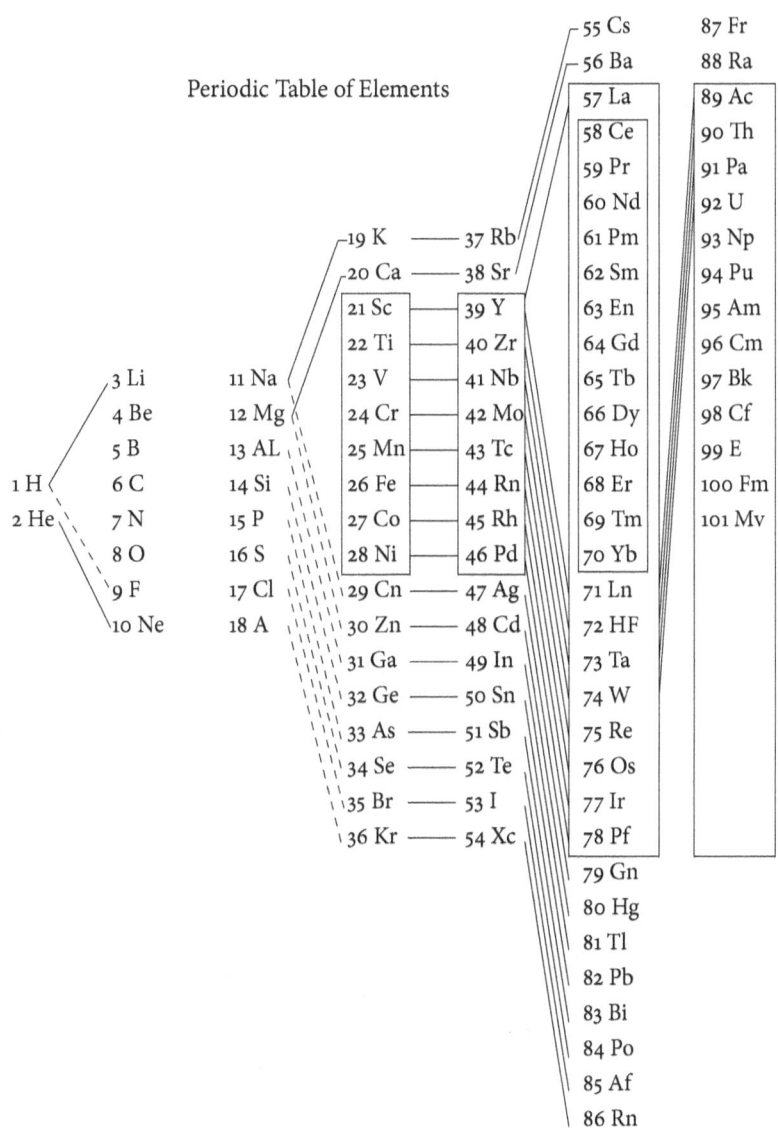

Periodic Table of Elements

Integral ratios and design

Levels within the atom

It was Niels Bohr who discovered in 1913 that it is the nucleus, which introduces system into the atom, that produces the extranuclear shells. The protons, neutrons, mesons, and so forth that make up the nucleus revolve by means of another, special energy level, as the chemist

Cosmic Purpose

Amount of orbital movement	1 iCell	Total angular	Number of nucleons inside the level	Number of nucleons up to that level
0	1 s	1/2	2	
				2
1	1 p	3/2⎤	4	
		1/2⎦	2	
				2
2		5/2⎤	6	
0		3/2⎥	4	
		1/2⎦	2	
				20
		7/2		
		↑↓		28
3	1 F	5/2⎤	6	
		3/2⎥	4	
		1/2⎦	2	
1	2 p	9/2⎦	10	
		↑↓		50
4	1 g	7/2⎤	8	
		5/2	6	
2	2 d	3/2	4	
0	3 s	1/2	2	
		11/2⎦	12	
		↑↓		82
5	1 H	9/2⎤	10	
		7/2	8	
3	2F	5/2	6	
		3/2	4	
		1/2	2	
1	3 p	13/2⎦	14	
		↑↓		126
6	1 i	11/2	12	

Levels within the nucleus of the atom

Maria Goeppert Mayer discovered at the University of Chicago sometime after 1905. She found that the bonding of elementary particles is never a matter of chance but always the result of fully selective arrangement.

Mayer not only argued that this arrangement takes place because of energy levels, but she further calculated that each elementary particle has its own rotation. According to Einstein's principle of relativity, all movement possesses mass; therefore, the claim that the elementary particles provided a level within the nucleus to rotate means that this rotation has at least a small influence on the mass of the elements. Mayer produced the table below. Her point regarding levels within the nucleus of the atom no doubt marked a major milestone in atomic physics. Still, more details were yet to be uncovered concerning the levels of atomic isotopes.

Some scholars likened the bonding of elementary particles within the nucleus of the atom to the combining of liquids, but as research on the atomic bond progressed, it was observed that the nucleus of the atom contained a "microcosm" within it. To begin with, there was no dismissing the fact that "light" was transformed into elementary particles, within which energy moved at a velocity approaching the speed of light. Thus, restoring the elementary particles would create lethal rays capable of killing hundreds of thousands of people in a second. The extreme speed of this movement made it easy to understand that, even in the case of the uranium atom, 138 protons and neutrons were packed within the narrow confines of the nucleus. They did not collide with one another but maintained their energy levels. According to the principle of relativity, as velocity increases, mass contracts and gravity is added. This suggested that the energy comprised of protons and neutrons would work at very high velocity and, once dispersed, produce lethal rays.

What we need to note here is that these subnuclear particles have orderly energy levels. Surely the time will come to fundamentally revise the hypothesis of H. G. Wells and Julian Huxley that life emerged through a "chance conjoining."

The integers and standards of isotopes

The presence of standardization in isotopes does not support claims like those of Wells that new things evolve by chance through natural selection. Detailed investigation of isotopes reveals that even when the number of neutrons or other particles has no particular order, once the number of protons is set, there is no t the slightest change in atomic number.

The number of electrons revolving externally is in exact relative proportion to the number of protons that form the nucleus. In light of this, the norms of selection latent in nature are of fixed determination and do not alter from one moment to the next. As nuclear physicists note, what chemistry usually refers to as the "magic numbers" 2, 8, 20, 50, 82, and 126 holds sway in the physics of the atomic nucleus as well.

Integral ratios in celestial bodies

The late Edwin Hubble, former Harvard professor and head of the Mount Wilson Observatory, discovered that the disposition of nebulae spread across the universe is not disorderly. He announced that two million light years away we run into an "island universe," where there lies a nebula measuring some one million, ten thousand light years in diameter, and that we do not run into another nebular realm without going more than two million light years beyond that.

The most interesting thing in the world of whole numbers is the fact that orbits within the solar system show proportions very similar to the internal structure of the atom. The integral ratios of distance that make up the nebulae and the sun are altogether amazing.

According to Cambridge professor Sir James Jeans, around 1772 Johann Elert Bode discovered the following integral ratios. We begin with

$$0\ 1\ 2\ 4\ 8\ 16\ 31\ 64\ 128\ldots$$

Now every number from the third in the series is the double of the preceding number. Tripling them, we get

$$0\ 3\ 6\ 12\ 24\ 48\ 96\ 192\ 384\ldots$$

If we next add 4 to each of these numbers, we have

$$4\ 7\ 10\ 16\ 28\ 52\ 100\ 196\ 388\ldots$$

Taking 10 as the base distance between the earth and the sun, the above proportions yield these approximations:

Mercury	3.9
Venus	7.2
Earth	10.0
Mars	15.2
Jupiter	52.0
Saturn	95.4

Uranus	191.9
Neptune	300.7

Bode proposed these laws prior to the discovery of Uranus, Neptune and the asteroids, but even after Uranus and the asteroids were identified, his law still held.

Following Jeans, the approximation for Uranus seems a bit too far off the mark and Mercury does not fit the general law. When we apply it to ratios of the distance between planets further away than the earth, and take 2 as the base unit of the distance between the sun and the earth, the distance of Mercury from the sun becomes ½. If we multiply this by 3 and add 4, we get 5½. Given that the actual ratio is 3.9, the gap is rather large.

But in my view, if we take the earth as the center, we have:

$$0 \quad 1 \quad 2 \quad 4 \quad 8 \quad 16 \quad 31 \quad 64 \quad 128$$
$$(N-2)^3, \; N, \; N^2, \; N^3, \; N^5, \; N^6, \; N^7$$

This means that the approximations are much closer to the alignment of electrons inside the atom than Jeans had thought. The electron cells unfold at $2N^2$ and $2(N-1)$, but the number of electrons is a multiple of 2. While Jeans states there is no completely clear theory here, it is curious that the structure of celestial bodies is so similar to that of the atom.

The law of integral ratios in music

The material world is relative, but Max Planck insisted that within this relativity are selective proportions with standards. The clearest proof is to be found in music. Sound frequencies differ widely from one thing to another. This means that when a number of sets of sound waves with different frequencies are drawn together, sets of "relative intervals" are produced. If these sound frequencies are then aligned in the following integral ratio, any one of these sets produces music with a complete octave:

$$24 \quad 27 \quad 30 \quad 32 \quad 36 \quad 40 \quad 45 \quad 48$$

It was Sir William Bragg who experimented with a set of cogged wheels of varying size driven by an electric motor to show that these integral ratios apply to musical intervals. When the wheel turns and one of the teeth hits the edge of a calling card, it gives off a put-put sound as a wave is sent through the air from the surface of the card. As the rotation of the wheel picks up speed, a succession of these waves creates a certain music—the faster the rotation, the higher the pitch.

There are eight such wheels, each with the following number of teeth:

Cosmic Purpose

24 27 30 32 36 40 45 48

Without any connection to the number of rotations made by the motor itself, the relationship among the sounds from each of the cogged wheels is that of the musical scale. The sound emitted from a unit with 48 teeth is always one octave higher than that of a unit with 24 teeth. In other words, eight notes is the interval determined at a ratio of 2:1. The relationship of a unit of 36 teeth to one with 24 is five notes, or a proportion of 2:3. Musical intervals therefore show a corresponding ratio.[4]

The standardization and discontinuity of integers

When we think of selection in the universe, the fact that the world of physical phenomena possesses a quality of discontinuity in respect to the continuity of wave motion is extremely significant. Discontinuity is also useful in the Morse code; without it, code could not be sent from one party to another. This discontinuity makes various combinations possible, and this in turn, through the purpose involved in all sorts of selection, enables the creation of a world with meaning.

When the meaning produced in such discontinuity is converted into integers, an interesting standard is set in place. The six dots of the Braille writing system can reproduce all twenty-six letters of the alphabet. Through the positioning of these six dots, the visually impaired can penetrate the world of meaning by the sense of touch.

Advertising posters on street corners with short sentences are actually arrangements of standardized discontinuous points. The various plays of the Japanese chess board and the movements of the stones on the *Go* board are performed by means of an infinite and apparently chance variety of movements that rely on a standardized discontinuity. To children the movement of the stones seems to take place more or less blindly, but to a master of the game the standardized activity of the black and white stones 316 (19×19) actually represent a goal-oriented world similar to the combat of great armies.

In this same vein, it is precisely through the standards of the integers of quantum mechanics exhibited in the world of physical phenomena that the presence of a complete finality in the universe can be demonstrated.

---

4. William Henry Bragg, *The World of Sound: Six Lectures Delivered before a Juvenile Auditory at the Royal Institution, Christmas 1919* (reprint: General Books, n.d.), 13–14.

The call number of objects

Just as the apparatus of automatic call numbers is able to handle telephone communications between millions of people in a large city, the universe with its integral standards makes use of discontinuity to create an infinite number of combinations. Byzantine mosaics combine angles and circles; the mechanical conventions governed by warp-and-woof in Gobelin tapestry and Japanese Nishijin weaving allow for the free expression of design in high-level artistry.

Television transmission also applies these integral ratios. Hence design is manifest in the midst of apparently random change as well as in the smooth running of machinery. Physical phenomena have numbers: their atomic numbers give us immediate and complete information on the number of electrons and protons in matter, and even, as Moseley discovered, a wave-frequency count. In much the same way as we can contact an individual by adding a private number to an area code, the universe is ordered through call numbers.

Design by chance

Chance is also transformed into design by using discontinuity in the realm of whole numbers. Prior to the discovery of the gene, Gregor Mendel's laws of inheritance were explained though coefficients of quadratic and polynomial equations, leading to the claim that inheritance was the result of chance combinations. This is because multi-dimensional equations appear to be the random result of multiplying linear equations again and again.

But let us not overlook everything that is well-ordered and law-regulated, where multiplication is *not* a matter of chance. There is also regulation and arrangement in the broad sense to be found. If it were all chance, the coefficient of multi-dimensional equations with probability would not take on a definite form. Even if one were to insist on seeing chance at work, it would have to be a *designed* chance, a *standardized* chance, a *limited* chance. Along with advances in contemporary research on genes, the selectivity at work in genes has become clear, with the result that the idea of "chance by design" has gained in precision. That is, a mutability approximating chance is present in the activity of genetic factors. But it turns out that this chance represents change close to the integral ratio of the coefficient of the equation of the possibility of creating a seedless watermelon. Severe change can lead to the apparent chance occurrence of a child born with a physical deformity; it can just as well lead to the birth of a genius. But this is *reasonable* chance, not completely *nonsensical* chance.

The coefficients of multi-dimensional equations expressed in integers help predict their roots. The graphic representations of coefficients opened up the world of calculus. In the same way, given the occurrence of integers appearing among the physical phenomena of the universe and then taking this as a coefficient in the cosmic equation, we hit upon the idea that at its base is a "root" that exists as a rational entity. And with the disclosure of these cosmic equations comes the conviction that the world of integers that carries it cannot be permanently blind either but is rather something that permeates it in a reasonable fashion. In this way the appearance of integers guides us to the world of selection.

Integers and design

As the study of quantum mechanics progressed, one fact that became clear is that the world of physical phenomena arose through integral multiples of quanta. Arthur Haas opens his book *The World of Atoms* by affirming that in the atomic world "the entities which take part in occurrences appear in perfect *accord* and the events themselves are revealed as governed by the most beautiful *whole-number harmony*."[5]

Haas is not alone in this view. Max Planck and Louis de Broglie made similar claims. Still, leaving aside the question of who agrees or disagrees with identifying the appearance of numerical formulae that include the integers of physical phenomena as foundational and praising it as highly as Haas does, no physicist today would deny these kinds of facts—which in itself is a rather startling.

Philosophically speaking, integers are rational numbers and the most readily understandable of all the calculations to issue from the human brain. To say the universe is governed by formulae that include whole numbers is to say that the universe is governed by a reason with a form similar to the human brain. This was pointed out by Hermann Weyl in works like *Mind and Nature* and *The Open World*.[6]

---

5. Arthur Haas, *The World of Atoms: Twelve Non-Mathematical Lectures* (New York: Van Norstrand, 1937), 2. Emphasis in original.

6. Hermann Weyl, *Mind and Nature: Selected Writings on Philosophy, Mathematics, and Physics* (Philadelphia: University of Pennsylvania Press, 1934); *The Open World: Three Lectures on the Metaphysical Implications of Science* (Woodbridge, CN: Ox Bow Press, 1989, orig. 1932).

If the world were ruled by chance, there would be no question of whole numbers being singled out to govern the world of physical phenomenon. Even so, irrational numbers and imaginary numbers still have a great deal to say. Irrational numbers are those like $\sqrt{3}$, $\sqrt{5}$, and $\sqrt{7}$, which can never be worked out; imaginary numbers are those that cannot even be conceived of, such as $\sqrt{-1}$.

Kyoto University professor of physics Yukawa Hideki has suggested that "imaginary numbers may well govern the radial vector of an electron."[7] He may be right. In the domain of discontinuities, Yukawa's argument is indeed plausible. But if we take the parts one by one as the norm, by and large standards of integral ratios apply. George Gamow observed that apart from isotopes, all the elements may be expressed as rational integers. Or again, in the interior of the atom, we may note that a fixed integer of 6.55×10^{-27}, or what is known as "Planck's constant," is recognized as the ultimate integer in the universe.

Dr. Naruse Masao, an authority on cogwheel research, has clearly shown that without an integral ratio a cogwheel is not complete, which obliges us to think that integers are necessary for things to revolve smoothly in the physical world. Indeed, without the appearance of integers it would not be possible for myriad subtle movements to proceed towards a goal.

From the minutest quantum in the field of thermodynamics to the celestial bodies that have evolved in the cosmos, there is no place that is not governed by integers. Once we see this, we understand why it is that crystals are built up geometrically according to simple ratios of whole numbers, or why the calories generated among molecules all show up in integral proportions. In other words, the universe we inhabit is one of integral geometry and integral mechanics. It is a realm to which purpose-oriented selection has been added and must therefore be thought of as emerging with a "design."

A note on the mathematics of music[8]

Western music uses what is known as the 12-tone equal temperament scale, which grew out of the need for vocal harmony. Tones are divided

---

7. See Yukawa Hideki 湯川秀樹, 『存在の理法』 [The law of existence] (Tokyo: Iwanami Shoten, 1943).

8. Nakamura Seiji 中村清二, 『自然と数理』 [Nature and mathematics] (Tokyo: Kokon Shobō, 1931), 191–3.

Cosmic Purpose

into twelve steps based on what is sensed between a beginning and an ending tone. That is, if we take the frequency of the beginning tone to be 1 and the ending tone to be 2, and calculate the 12th root, we come to the average musical interval of one step. If the beginning tone begins from $(^{12}\sqrt{2})^0 = 1$, it is followed by $(^{12}\sqrt{2})^1 = 1$, $(^{12}\sqrt{2})^2 = 1$, and so on, proceeding through the twelve steps $2^{\frac{0}{12}}\ 2^{\frac{1}{12}}\ 2^{\frac{2}{12}}\ \ldots\ 2^{\frac{11}{12}}$ until it reaches the end tone. One step of the musical interval, $2^{\frac{1}{12}}$, is called a half-tone interval and its double, $2^{\frac{2}{12}}$, a full-tone interval. If we then compare the 12-tone equal temperament to the steps in the natural scale, *do–re* and *re–mi* represent full-tone intervals, *mi–fa* is a half-tone interval, and *fa, so, la,* and *ti* are full-tone intervals. The ending tone between *ti* and *do* is seen as a half-tone interval with no equivalent difference.

Japan's ancient court music or *gagaku*, like the ancient music of China, also set up a scale of 12 steps with intervals between the beginning and ending tone. Unlike Western music, however, the musical intervals are constructed according to a law of "8 rising and 6 descending" tones. This means that the eighth tone higher than a given tone has a frequency of 2:3, while in the reverse progression from a higher to a lower tone, the sixth tone has a frequency of 4:3. As illustrated in the diagram below, multiplying the beginning tone or *ichikotsu*, by $\frac{3}{2}$ gives the eighth tone or *ōshiki*, which, if multiplied by $\frac{3}{4}$, gives the sixth tone or *hyōjō*. This was the original theory of Japan and China, though musical performance did not necessarily follow it. According to the theory, the ending tone of *ichikotsu* would not be 2 but 2.0273; in practice, of course, it was rounded off to 2.

If Western music gives the true frequency of *la* (3) at 435 per second, in Japanese *gagaku* the frequency of *ichikotsu* is given at 290 per second.

In late seventeenth-century Japan, Nakane Genkei worked on the 12-tone equal temperament scale to come up with the *gagaku* scale. China's theory followed somewhat later, with Zhu Zaiyu in the Ming period, though it was not actually used in practice.

The mathematical structure of botanical phyllotaxy[9]

The famous Schimper-Braun series has become well known. Divergences among leaves arranged spirally around a mature stem can be expressed in a series of fractions: $\frac{1}{2}\ \frac{1}{3}\ \frac{2}{5}\ \frac{3}{8}\ \frac{5}{13}\ \frac{8}{21}\ \frac{13}{54}\ldots$. This makes use of the Fibonacci sequence to illustrate that the sum of any two continuous numerators or

---

9. Fujita Tetsuo 藤田哲夫, 『植物の器官形成』 [The formation of plant organs] (Tokyo: Kawade Shobō, 1948), 35–41.

12-Tone Equal Temperament Scale	Natural Scale		Twelve-Tone *Gagaku* Scale	
$2^{\frac{12}{12}} = 2{,}0000$	do	2	壹越 ichikotsu	$\frac{531414}{262142} = 2{,}0273$
$2^{\frac{11}{12}} = 1{,}8877$	ti	$\frac{15}{8} = 1{,}8750$	上無 kamimu	$\frac{243}{128} = 1{,}8984$
$2^{\frac{10}{12}} = 1{,}7818$			神仙 shinsen	$\frac{59049}{32768} = 1{,}8020$
$2^{\frac{9}{12}} = 1{,}6818$	la	$\frac{5}{3} = 1{,}6667$	盤渉 bannshiki	$\frac{27}{16} = 1{,}6875$
$2^{\frac{8}{12}} = 1{,}5874$			鸞鏡 rankei	$\frac{6561}{6096} = 1{,}6018$
$2^{\frac{7}{12}} = 1{,}4983$	so	$\frac{3}{2} = 1{,}5000$	黄鐘 ōshiki	$\frac{729}{512} = 1{,}4238$
$2^{\frac{6}{12}} = 1{,}4142$			鳧鐘 fushō	$\frac{729}{512} = 1{,}4238$
$2^{\frac{5}{12}} = 1{,}3348$	fa	$\frac{4}{3} = 1{,}3333$	双調 sōjō	$\frac{177147}{131072} = 1{,}3515$
$2^{\frac{4}{12}} = 1{,}2599$	mi	$\frac{5}{4} = 1{,}1250$	下無 shimomu	$\frac{81}{64} = 1{,}2656$
$2^{\frac{3}{12}} = 1{,}1892$			勝絶 shōzetsu	$\frac{19683}{16384} = 1{,}2013$
$2^{\frac{2}{12}} = 1{,}1225$	re	$\frac{9}{8} = 1{,}1250$	平調 hyōjō	$\frac{9}{8} = 1{,}1250$
$2^{\frac{1}{12}} = 1{,}0594$			断金 tangin	$\frac{2187}{2048} = 1{,}0679$
$2^{\frac{0}{12}} = 1{,}0000$	do	1	壹越 ichikotsu	1

denominators is equal to that of the following numerator or denominator. When these values are applied to actual plants, the numerator marks the number of times the spiral winds around the stem in the ascent from a given leaf to that immediately above it; and the denominator marks the number of intervals into which the ascent was divided by successive leafs. Take the fraction $\frac{1}{3}$, for example. Immediately above leaf number 1 is leaf number 4, and between them the spiral winds about the stem 1 time. Similarly, the fraction $\frac{2}{5}$ indicates that the leaf directly above leaf number 1 is leaf number 6. Between the two are leafs 2, 3, 4, and 5, up to which point the spiral winds about the stem 2 times. Because these Schimper-Braun values express divergencies in circumference by means of fractions, when they are converted to angles, each of the fractions is multiplied by 360°

degrees, so that $\frac{1}{3}$ gives 120° and $\frac{2}{5}$ gives 144°. In the case of a divergence of $\frac{2}{5}$ the next leafs

$$1, 6, 11, 16$$
$$4, 9, 14$$
$$2, 7, 12, 17$$
$$3, 8, 13, 18$$

are aligned vertically and in a straight line. These lines are referred to as columns, so that in this case the leafs line up vertically in 5 lines, giving 5 columns. In the same way, with a divergence of $\frac{3}{8}$, there are 8 columns and with $\frac{5}{13}$, 13. Accordingly, in the Schimper-Braun values the denominator indicates the number of columns and at the same time the sum of the number of cycles, whereas the numerator shows the number of spirals intersecting with the next level below.

This way of demonstrating divergences is not actually the same as that found in the arrangement of the original plant in the process of growth. It calculates from leaves on a mature stem, leaving a certain discrepancy between the two. Vascular bundles on the stem, when they first form, are not aligned vertically, many of them running in a line alongside one group of diagonally slanted intersections. For example, in a ratio of 3:5 they run along 8 lines, and in the case of 5:8, 13 lines.

As the stem stretches out, the system surrounding the stem grows more slowly, so that as the vascular bundles stretch they do so resisting the surrounding system. As a result, vascular bundles that are slanted get straightened out in a line and return to the very shape they had before. Along with this, the stem as a whole is twisted and the divergence worsens. For example, because in an item with a diagonally slanted intersection of 2:3 the bundles normally run alongside the fifth lower row, when it is straightened out, the divergence becomes $\frac{2}{5}$. In the same way, an item with a 3:5 ratio shows a divergence of $\frac{3}{8}$. In this manner, divergence after extension differs slightly from the divergence during growth.

The employment of the Schimper-Braun series is very closely related to this. Given the Fibonacci series

$$1, 2, 3, 5, 8, 13, 21, 34, \ldots m, n$$

the proportion between two items is to be sought in the items bordering it on either side:

$$\frac{1}{2} \quad \frac{1}{3} \quad \frac{2}{5} \quad \frac{3}{8} \quad \frac{5}{13} \quad \frac{8}{21} \quad \frac{13}{54} \quad \cdots \quad \frac{m}{n}$$

At first the difference is large, as in $\frac{1}{2}$ and $\frac{1}{3}$, but as the series progresses

Standardization in Natural Selection

it grows smaller, wandering between $\frac{1}{2}$, and $\frac{2}{3}$, gradually approaching a fixed value, whose limit value is expressed as:

$$\frac{1}{2} = \frac{1}{1+1},$$

$$\frac{2}{3} = \frac{1}{\frac{3}{2}} = \frac{1}{1+\frac{1}{2}} = \frac{1}{1+\frac{1}{1+1}},$$

$$\frac{3}{5} = \frac{1}{\frac{5}{3}} = \frac{1}{1+\frac{2}{3}} = \frac{1}{1+1}1+\frac{1}{1+\frac{1}{1+1}}$$

$$\cdots\cdots\cdots\cdots\cdots\cdots\cdots\cdots,$$

$$\cdots\cdots\cdots\cdots\cdots\cdots\cdots\cdots.$$

Now if we set the limit value to x,

$$\mathrm{Lim}\,\frac{m}{n} = x$$

$$= \cfrac{1}{1+\cfrac{1}{+1\cfrac{1}{1+\cfrac{1}{1+}}}}\cdots\cdots$$

$$\therefore x = \frac{1}{1+x}$$

$$x^2 + x - 1 = 0$$

$$\therefore x = \frac{\sqrt{5}-1}{2}$$

$$= 0.6480340$$

This value is known as the "golden ratio." In geometry, it occurs when point c divides a line ab such that a and b are are related $\frac{ac}{bc} = \frac{ab}{ac}$. That is, if $ab=1$ and $ac=y$, then

$$\frac{ac}{bc} = \frac{ab}{ac},$$

$$\frac{y}{1-y} = \frac{1}{y},$$

$$y^2 + y - 1 = 0$$

$$\therefore y = 0.6180340$$

The rectangle formed by the golden ratio of these two proportions is the most aesthetically pleasing. Taking the Schimper-Braun divergences

$$\frac{1}{2}\ \frac{1}{3}\ \frac{2}{5}\ \frac{3}{8}\ \frac{5}{13}\ \frac{8}{21}\ \cdots\cdots\frac{p}{q}$$

and seeking out its ultimate value, we have:

$$\frac{1}{2} = \frac{1}{1=1},$$

$$\frac{1}{3} = \frac{1}{2=1},$$

$$\frac{2}{5} = \frac{1}{\frac{5}{2}},\ \frac{1}{2+\frac{1}{2}} = \frac{1}{2 = \frac{1}{1=1}}$$

Cosmic Purpose

$$\frac{3}{8} = \frac{1}{\frac{8}{3}} = \frac{1}{2+\frac{2}{3}} = \frac{1}{2 = \frac{1}{1\frac{1}{1+1}}},$$

$$\frac{5}{13} = \frac{1}{\frac{13}{5}} = \frac{1}{2+\frac{3}{5}} = \frac{1}{2+\frac{1}{1+\frac{1}{1+1}}}$$

$$\cdots\cdots\cdots\cdots\cdots\cdots\cdots\cdots\cdots\cdots ,$$
$$\cdots\cdots\cdots\cdots\cdots\cdots\cdots\cdots\cdots\cdots ;$$

$$\therefore \operatorname{Lim} \frac{p}{q} = \frac{1}{2+x}$$
$$= \frac{1-x}{(2+x)(1-x)}$$
$$= \frac{1-x}{2-(x+x^2)}$$

Therefore,

$$x^2 + x = 1 ,$$

$$\therefore \operatorname{Lim} \frac{p}{q} = 1 - x$$

$$= \frac{3-\sqrt{5}}{2}$$

$$= 0.3819660$$

Accordingly, if we transform this into an angle (discarding numbers 3 or less for the sake of convenience), we have $360° \times 0.38 = 137°30'28"$. This angle is thus the smallest angle that can be made by dividing $360°$ according to the golden ratio.

Yukawa Hideki's theory of the elements and standards in the universe

Yukawa's "law of existence"

On 15 July 1943, Yukawa Hideki, who left a world-class impression on the field of physics through his discovery of the meson, published a book entitled *The Law of Existence* in which he set forth an interesting idea of finality in the physical world. In the opening chapter he wrote:

> The laws of classical theory provide cause-effect relationships and at the same time, from another side, we can say that grounds have been provided for thinking that nature follows a path preestablished by a determined purpose.[10]

These words display a concern that set Yukawa apart from physicists up

---

10. Yukawa Hideki 湯川秀樹, 『存在の理法: 現代物理學の根本問題』 [The law of existence: Fundamental problems of contemporary physics] (Tokyo: Iwanami, 1943).

to that time in that he identified the world of cause and effect as a world of *finality*. It is generally reasoned that in the law of cause and effect, each cause has one effect and that a cause sets determined limits on possibility. For that reason, a goal allows for only one selection.[11] Even so, Yukawa has his doubts about finality in the universe when it comes to the minutest activities.[12]

When we enter the microscopic world that structures matter, we encounter facts that run directly counter to the macroscopic world: if a position is decided on, velocity becomes uncertain, and if velocity is decided on, position cannot be determined. In pointing this out, Yukawa argues that elementary particles even lack form.[13]

Classifying elementary particles as electrons, photons, protons, positrons, neutrinos, mesons, and neutrons, Yukawa claims that in order to explain the mechanics of the particles it is necessary to take a step beyond the wave mechanics and quantum mechanics of de Broglie, Schrödinger, and Heisenberg. He observes that "energy" forms a field by means of configuration and constitutes matter in a discontinuous manner. When "energy" is bound to configurational space, a realm of selection is the inevitable result, as is clearly known in crystallography.

Furthermore, Yukawa considers the concept of "time" extremely important in the realm of quantum mechanics and finds great meaning in the "finite degree of freedom" in matter. That is, he holds that through restriction, possibility becomes probability. He has also noted that whereas the realm of wave mechanics implies a continuous world, in the case of passive movement, it becomes transcausal and discontinuous.[14]

The evolution of the concept of space

In the second chapter of his book, Yukawa turns to modes of the laws of nature. He argues throughout that alongside the evolution and development of human knowledge, the meaning of *space* also evolved and grew—from Euclidean space to Minkowski space to Einsteinian space. He recognizes that changes in the subjective structure of the idea led to the growth of laws.

While I do not accept that the laws of nature themselves are evolving

---

11. *Ibid.*, 81.
12. *Ibid.*, 29.
13. *Ibid.*, 189.
14. *Ibid.*, 38.

in an objective sense, we need to think of laws evolving in terms of human subjectivity. Carried to an extreme, Yukawa's hypothesis undoes the truth of laws. Einstein was extremely cautious in this regard. Although he theorizes about the relativity of matter, he believes that the laws of nature are absolute. Objectively speaking, he is right.

"We may consider the laws of nature to restrict the infinity of possibilities. There is always a kind of 'law of selection.'"[15] Yukawa believes that it was the laws of the cosmos that introduced cause and effect by selecting from among an infinity of possibilities. I consider this opinion to be correct. Yukawa holds, however, that the world of cause and effect is continuous and absolute, which is the world of classical physics.

The realm of elementary particles is a discontinuous world with a "finite degree of freedom." Not only is it governed by integers, but also, as Yukawa goes on to hypothesize, imaginary numbers have their place and a velocity faster than the speed of light is a ruling force as well. In other words, he supposes that a velocity greater than light and mathematically imaginary numbers are at work within the diameter of elementary particles:

> Therefore, when we think of a situation in which cause and effect cannot be separated—as Taketani Mitsuo has pointed out—this probably means that we are drawn closer to a teleological point of view. At the level of present-day science, causal and teleological views are seen as opposing one another. But the more deeply we advance into the recesses of nature, does not the very presence of a kind of "law of existence" seem to dismantle both the causal and the teleological viewpoints? If we stop short of this, are we not left with a sense of inconsistency in which some of the phenomenon of nature seem to be causal and others purposive?[16]

Thus, even though Yukawa harbors doubts about teleology, at the same time, like Hermann Weyl, he holds to a belief in the developmental nature of the universe:

> The so-called classical physics of the nineteenth century was complete in itself. Like mathematics, it was self-contained. It mattered not if it happened to contradict what lay outside of it. It had no need to reflect on things like religion, art, or history. Because it was closed, it left no room for such things. But when we come to the present day,

---

15. *Ibid.*, 85.
16. *Ibid.*, 78.

physics is not so clearly self-contained and there are things that do not make sense if looked at only through the eyes of physics. From a broader perspective, this is for the best. Nineteenth-century physics was closed. Physics today is open. Scholarship as such always tends to close itself in on itself. We have to make the effort to open it up. Once the window is thrown open, light comes in from the outside.

Hence, physics today is—how shall we put it?—a bit of a jumble. In quantum mechanics, for example, the question of observation and the like is somewhat muddled. There are times we simply cannot make sense of things. But it is precisely for this reason that a way is left open to understand all sorts of things like life or history or religion. It may seem to be complete in itself, but it is not.[17]

De Broglie, at the end of his *Matter and Light*, expresses views more or less the same as Yukawa's. But where would the ground lie in physics for acknowledging a purposive view of the universe outside of wave theory? This is something we would like to know. In a second appendix to *Law and Existence* where he takes up time and causality, Yukawa notes the contrasting explanations: Bohr holds that the world of life complements the world of physical science; Jordan and others argue for the possibility of free will and teleology in the realm of elementary particles.

The direction of a priori selection

Why does electricity run in a plus direction within a dexiotropic helical coil? Is something determining this a priori selection? A magnetic force is said to move vertically to the movement of the electricity. What decides this? There are countless cases of just such a priori selection in the universe. The earth spins from west to east, and has done so unchanged for billions of years. How did this a priori principle of selection come about?

The prearrangement of atomic valence, the principles of chemical combination, the determination of molecular equations, the selectivity in catalysis, the startling selectivity in organic oxygen—how many tens of thousands of examples would we find just by taking the selection we find at work in the ordinary numerical constants that appear in the realms of physics, chemistry, and biology!

A close look at the nature of a priori selection leads us to the following:

17. *Ibid.*, 156–7.

1. All change is subject to a priori limits. This is called the principle of *a priori selection*.
2. In cases where a priori selection occasions a scientific response, it possesses an *a priori probability*.
3. In shutting down selection, a priori probability introduces the possibility of law as well as mathematical calculation. Geometrical and algebraic theorems are established by means of the a priori probabilities provided by such a priori selection.
4. Physical selection principles and a priori selection principles form the structural foundations of physics prior to the transformation of the universe into matter through light, electricity, and magnetism. This may be thought of as *primary selection*.
5. Chemical selection principles and a priori selection principles, by means of the secondary organization seen in matter, brings into existence a physical universe that enables a mechanistic organization from which a priori probability may be expected. This may be thought of as *secondary selection*.
6. A *tertiary physiological selection* principle develops in the form of solids, liquids, gases, electricity, magnetism, and radiation in which a priori selection possesses a degree of freedom, and organizes the organic world through a selective activity that brings into existence a counteraction, enriching what is undergoing development. This leads to the appearance of the "phenomenon of life" in living entities with its "orientation toward unified conditions" and "tendency toward the regulation of materials."

The higher-level meaning of a tertiary a priori selection principle

The a priori selection that touched off the phenomenon of life is not the sort of disorderly process that can be thought of as a blind sifting in nature. At least seven high-level choices were in place beforehand, namely selection involved with:

1. centrality of laws
2. accumulation of functional mechanisms
3. self-regulation in response to antagonism
4. orientation by means of continuous enzyme reactivity
5. adaptational intentionality in coping with changing circumstances
6. homeostasis that persists through open-ended instability
7. the tendency to unification

With the phenomenon of life as our axis, nothing at all stands in the way of treating the development from the primary to the tertiary level of a priori selection mechanically. Yet because these mechanisms represent a selection of direction running through the three levels, we may also think in terms of selection with a latent *finality*.

The ancient Greek philosopher Aristotle, in his treatment of organic bodies, prescribed a finality to the appointed tasks of parts vis-à-vis the whole. Aristotle did not hesitate to acknowledge a purpose in dynamic and chemical selection aimed at the conditions of life, inasmuch as primary dynamic a priori selection, secondary chemical a prior selection, and tertiary physiological a priori selection all contribute as parts to the whole in the emergence of life.

We may thus suppose that even though it lacks a conscious finality, life lays the ground for finality through inclination and orientation. In this connection it is worth mentioning Wilhelm Pfeffer, a disciple of the pioneering plant physiologist Julius von Sachs.

The inclination to a steady state in the universe

Since the invention of the atomic bomb, the instability of the atom has been widely researched. The public may not pay much attention to the nature and degree of this instability, but in order for life to be generated on earth, the atoms that constitute the universe would not have been able to bring life to earth if they were unstable. For this to happen, a world with a steady state is required.

Norman Feather, lecturer in natural philosophy at Edinburgh University, takes this point up in a detailed exposition of the laws of stability that pervade the atom. Basing his conclusions mainly on recent experiments, he points out the laws of selection at work in the natural world. We may recall that Maria Mayer drew attention to the energy level in the nucleus of the atom by relating the stability of the nucleus to a "magic number." Feather goes on to note that the atomic number (or Z) of tin, 50, also has 10 isotopes, xenon (Z=54) has 9 isotopes; and cadmium (Z=48) and tellurium (Z=52) each have 8 isotopes. Among atoms with a low atomic number, only calcium (Z=20) has 6 isotopes. This gives us the "magic numbers" 8, 20, 50, and 82. William D. Harkins had observed in 1949 that protons and neutrons have N×4 as their nucleus, as in helium, which fits these "magic numbers."

After careful study of neutrons, the brilliant Italian physicist Enrico Fermi has shown us that the nucleus of the atom is composed of six kinds of "elementary particles": protons, electrons, positrons, neutrons, mesons, and neutrinos. Thus if conditions are even slightly off, instability sets in and mode of being breaks down. Thus when the protons and neutrons are strongly bonded and when the alpha particle (α) is destroyed, helium (He, Z=2) emerges. The fact that carbon (Z=6) whose atomic weight is 12, $^{-12}_{6}\text{C}$, or oxygen $^{-16}_{8}\text{O}$, magnesium $^{12}_{24}\text{MG}$, silicon $^{28}_{14}\text{SI}$, calcium $^{40}_{20}\text{CA}$, are the most commonly found on earth is due to the proper quantity of their "magic numbers."[18]

By reversing the algorithm, it is easy to deduce why zirconium (Z=40) and promethium (Z=61) are not to be found on earth.

The moon and the earth's stability

Henry Norris Russell has called the moon the "sanitary engineer" of the earth. Without the moon there would be no tides, the seas would grow turbid, and unclean things on the earth would not be cleansed by the swelling of the sea. But the great task of the moon is to provide stability to the revolution of the earth. It is thought that when the moon was close to the earth, around 120,000 miles, the changing of the tides was intense and conditions on the surface of the planet were too unstable to allow living things to inhabit it. About 500 million years ago the moon is thought to have withdrawn to its current distance and be nailed down there, after which the evolution of life began.[19] Harold Jeffreys thinks that if we go back 50 billion years, the moon drew closer to the earth and, like the rings around Saturn, clouds of cosmic dust formed around it, with the result that the tides stopped ebbing and flowing.

We are thus driven to think that the moon contributed greatly to the turning of the earth every twenty-four hours and the annual cycle of 365 days. As simple as it sounds, the remarkable work of the moon enables the temperature on the surface of the earth to lower in the transition from day to night.

Stability in local super-galaxies

Prior to 1543 when Copernicus proclaimed his theory of the earth's movement, astronomers considered the earth to lie at the center of the universe. Swedenborg, Thomas Wright, Immanuel Kant, Jean Lambert, and

---

18. Norman Feather, *Nuclear Stability Rules* (Cambridge: Cambridge University Press, 1952) 16–26.

19. Ralph B. Baldwin, *Face of the Moon* (Chicago: University of Chicago Press, 1949), 180.

others went so far as to suppose that the real universe was formed by the galaxy of the Milky Way, with the sun positioned along it.

In 1785, prior to the French Revolution, William Herschel, astronomer to King George III of England, held that the universe was round and had the sun at its center. This first model of the galaxy was later abandoned. In 1900 the Dutch astronomer Jacobus Kapteyn selected 206 locations out of the heavens to study the number, light intensity, color, and spectrum of the stars, arriving in 1922 at the same results as Herschel. Kapteyn's universe was a circle a mere 40,000 lights years in diameter. He believed the solar system to be positioned at the center with a sparser number of stars as one moved to the periphery. His views were, of course, subsequently understood to be mistaken.

At the time of the First World War, Harlow Shapley, a young astronomer at the Mount Wilson Observatory, began to work on variable stars in the Cepheid constellation. Taking their variations as a clue, he tried to discern the structure of the universe. It was then that he struck on the idea that the point at which stars come together in the Milky Way coincides with the center of the galaxy. He went on to estimate the diameter of the galaxy to be 300 thousand light years. When he later noticed that light waves from the nebulae that fill the universe are absorbed, he revised his figure downwards by 100 thousand light years.

Subsequently, in 1927, Bertil Lindblad of Sweden and Jan Oort of Holland proved that our galaxy rotated, which led them to roughly the same structure as Shapley. In recent years it has been observed that the galaxy is a vortex of nebulae and research continues in that direction.

Whether there are island universes outside of the Milky Way or not was a matter of debate, but in 1917 George Willis Ritchey of the Mount Wilson Observatory discovered a new star or "nova" from an explosion in the Andromeda nebula. Following this lead, the Director of the Observatory, Edwin P. Hubble, discovered the presence of

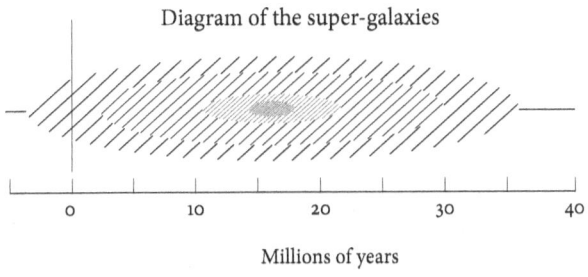

Diagram of the super-galaxies

Millions of years

Cosmic Purpose

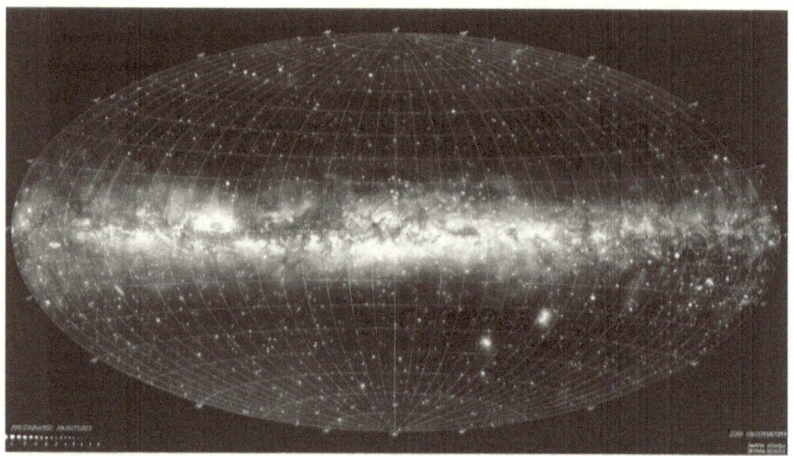

A mock-up of all the stars in the heavens with the Milky Way at the center
(Prepared by Martin and Tatjana Kesküla under the guidance of
Knut Lundmark, director at the Lund University Observatory in Sweden, 1955)

variable Cepheid stars within the Andromeda nebula, which eventually helped him understand that the Andromeda constellation lay outside our galaxy. In his famous study *The Realm of the Nebulae*, Hubble reckoned that island universes are galaxies that generally conform to rules and are scattered here and there with about two million light years between them. He conceived of the nebulae of these local galaxies as more or less shaped like a lens one million light years in diameter and consisting of clusters of stars like the sun.

With the recent development of the electromagnetic telescope, Hubble's hypothesis has been corrected. In the July 1954 issue of *Scientific American*, Gérard de Vaucouleurs reported that it was proper to consider the group of several hundred galaxies which includes the Milky Way a "local super galaxy" with the galaxy of the Virgo cluster at the center. Around 1923 the amateur astronomer J. H. Reynolds observed the presence of a single cluster crossing the vortex ring of the Milky Way at an angle of 100 degrees. He left the matter there without looking into it sufficiently. In 1932, Harvard University was able to confirm Reynolds's findings while examining a diagram he had prepared of 1,200 luminous local galaxies.

In 1953 de Vaucouleurs made an announcement that this super local galaxy was 40 million light years in diameter with a thickness at the center of 1 million light years. That same year J. D. Kraus of Ohio State University and his collaborators, along with Manchester University's R. Hanbury Brown and C. Hazard, employed an electromagnetic telescope to

Standardization in Natural Selection

The Andromeda constellation

locate at 13° latitude and 5° longitude the north pole of a group of stars in a local super galaxy. That point stood at the equator of a galaxy that includes the greatest concentration of star clusters visible from the northern hemisphere of earth.

There are several such local super galaxies comprising multiple local galaxies in the vicinity of the Milky Way. The Hydra and Pavo-Indus constellations are a southern local super galaxy that opens up 50° above the angle at which those constellations are clearly visible from the southern hemisphere. The number of still more distant galaxies being discovered keeps on growing. But this theory is in need of revision.

Prior to World War II, with the publication of Hubble's research it was thought that Sagittarius, which arose at the left shoulder of Scorpio as it arose in the southern sky at 8:00 p.m. in the month of August, was the center of a super local galaxy belonging to the Milky Way.

According to the hypothesis proposed by Vaucouleurs, there are several super local galaxies that are arranged in fours inner layers like an onion, and the Milky Way is stuck at the outermost ring. Harvard's Shapley had previously thought that the various star clusters within the Milky Way—of which ninety-five had been discovered—function like the ball bearings fitted around a bicycle shaft and that such a system was needed to insure stability in the universe.[20]

In Newtonian mechanics, the structure of the universe was that of a stable machine. Einstein's principle of relativity arrived in the early years of the twentieth century and around the end of 1924, De Broglie introduced wave mechanics, setting the way for scientists like De Sitter of Holland, Georges Lemaître of France, and James Jeans and Arthur Edding-

---

20. Araki Toshima 荒木俊馬,『銀河系』[The Milky Way] vol. 3/8 of『天文宇宙物理学総論』[General theories of astronomical cosmophysics:] (Tokyo: Cosmophysics Research Association, 1949), 141–81.

ton of England to advance theories of an expanding universe. Hubble attempted to confirm the theory with a telescope. Making use of the Doppler effect of physics and reckoning that distant light rays would tend towards violet on the light spectrum and those closer would tend towards red, he tried to demonstrate that the universe is expanding.

Meantime, Hermann Bondi of Cambridge University has proposed the hypothesis that because plus and minus electrons are invariably created through light rays, the universe has a steady-state stability.[21]

The stability of the elliptical path

It was Kepler who drew our attention to the stability of the solar system, whose elliptical path he reckoned as firmly settled. Because the heat of the sun, which provides each point on the ellipsis with a distance based on two foci, also depends on atomic explosions, this means that there is no fear of it simply burning out the way oxygen does in the atmosphere. Nevertheless, Gamow thought that as the amount of hydrogen in the sun diminishes, its bonding power would weaken, its surface area increase, and the quantity of heat falling to the face of the earth would rise.

Three momentous changes in our view of the universe

Around the time Charles Darwin published *On the Origin of Species* (1859), chemistry in general had not made any progress, and even physics had not yet had so much as a glimpse of wave mechanics. Consequently, the basic ideas he formulated in his theory of evolution followed three current hypotheses: Ludwig Büchner's materialism, the mechanism of Hermann von Helmholtz based on a physiological science that mainly involved the "imperfections of the eye," and the "theory of population" of Thomas Malthus, all of which suggested casualism.

Over the next hundred years outstanding scientists made advances in the fields of physics, chemistry, and biology. In physics, the pioneering work of James Clerk Maxwell on electromagnetism finally led to an electrical approach to matter that eroded confidence in the idea of the "indestructibility of matter" subscribed to in Darwin's time. In our own day, no one any longer doubts the author of *Matter and Light*, the French genius De Broglie, when he says that light is the substance of matter.

Along with advances in nuclear physics, Ernest O. Lawrence began by

---

21. Hermann Bondi, *Cosmology* (Cambridge: Cambridge University Press, 1952), 140–56.

conducting successful experiments in creating matter with light,[22] and by the end of 1924 De Broglie had actual proof endorsing the theory that light waves can become matter.[23] As a result, the theory of creation that Darwin and many evolutionists had laughed at saw a revival in astrophysics. In February of 1949, Sir Fred Hoyle of St. John's College of Cambridge published an essay in *Nature* on "Continuous Creation," which he later expanded on in a volume entitled *The Nature of the Universe*.[24]

Gamow, who made a great contribution to the invention of the hydrogen bomb, published a book entitled *The Creation of the Universe*, in which he began from a nuclear physics perspective completely different from that of previous theories of the "evolution of the universe"— a theistic (or as he called it, Augustinian) standpoint from which to work out a new theory of the creation of the universe.[25] These scientists had come to a point where advances in the development of nuclear physics, resulting in the manufacture of an atomic bomb and the peaceful manufacture of atomic reactors, had provided them with theoretical support for their views. But with the inauguration of De Broglie's wave mechanics, astronomers from around the world began to advocate the new cosmology of an "expanding universe." Arguments were put forward by De Sitter, Lemaître, Eddington, and Jeans,[26] among others. The actual measurements taken by Hubble provided confirmation.[27]

One upshot of the theory of an expanding universe was the idea that the stars in the cosmos had begun as light waves and had an end, but since nuclear physics in the 1930s was insufficiently advanced, courage was lacking to carry it over to a theory of the "creation of the universe."

Before World War I the French philosopher Henri Bergson, basing his thinking on theories of mutation, began to propose a philosophy of cre-

---

22. Ernest O. Lawrence, "The Physics of High Voltage," *Science in Progress* 4 (1945).

23. Louis De Broglie, *Selected Papers on Wave Mechanics* (London: Blackie & Son, 1928).

24. Fred Hoyle, *The Nature of the Universe* (New York: Harper, 1950). A Japanese translation by Suzuki Keishin appeared the following year.

25. George Gamow, *The Creation of the Universe* (New York: Viking, 1952).

26. Willem de Sitter, *Kosmos* (Cambridge, MA.: Harvard University Press, 1932); Georges Lemaître, *The Primeval Atom: An Essay on Cosmogony* (New York: Van Norstrand, 1950); Arthur Eddington, *The Expanding Universe* (New York: Macmillan, 1933); Sir James Jeans, *The New Background of Science* (New York: Macmillan, 1933).

27. Edwin Hubble, *The Realm of the Nebulae* (New Haven, CT: Yale University Press, 1936).

ative evolution[28] which, together with experimental advances in nuclear physics, occasioned fundamental corrections in Darwin's theories.

The dialectical materialism and materialistic views of history of Marx and Engels were expounded roughly at the same time as Darwin was writing and while atomic physics was still in its infancy. In later years the presence of a selection principle in the material world would be discovered and recognized as governing the atomic realm, something that did not occur to Marx and Darwin, not to mention the older critical philosophy of Kant.

High-level selection principles governing elementary particles

There are six types of super high level selection principles governing the realm of elementary particles. They show up in:

1. crystallography;
2. electrical activity, particularly spectrochemistry;
3. nuclear concentration;
4. polymer chemistry;
5. atomic valences in Mendeleev's periodical table; and
6. Avogadro's law in gas dynamics.

I defer to the following specialized writings for detailed information:

1. On the astonishing selectivity present in crystallography in 6 forms, 32 modes, and 230 types, the work of Linus Pauling is admirable.[29]
2. Regarding the wave lengths shown by electrical activity in spectrochemistry and their significance for quantum mechanics, Niels Bohr's analysis was epoch-making, and the reader is referred to his work.[30] The logic of selection in electrical paths was brought to my attention by Sommerfeld, who observed how an authoritative selection principle is at work in the laws of physical change.[31] To ordinary common sense, it would appear that when electrons get excited and distance themselves far from the nucleus of the atom, even if the distance is not measurable as an integral multiple, they would remain stable. This is not in fact the case. There is stability where there are "integral multiples," and surprisingly, even when they shift

---

28. Henri Bergson, *Creative Evolution* (London: Macmillan, 1911).

29. Linus Pauling, *The Nature of the Chemical Bond and the Structure of Molecules and Crystals:* (Ithaca, NY: Cornell University Press, 1940).

30. Niels Bohr, *The Theory of Spectra and Atomic Constitution: Three Essays* (Cambridge: Cambridge University Press, 1922).

31. Arnold Sommerfeld, *Three Lectures on Atomic Physics* (London: Methuen, 1926).

within the identical frame of energy levels, a strict law governs the selection of their path.
3. The concentration in the nucleus of the atom is not chaotic. As Maria Mayer has shown, it has the same energy level as the selection principle at work in electrons. (Of course, the theory has not yet been clearly applied to research on isotopes.)
4. We may consider things like the regulation of integral ratios appearing in protein polymer chemistry as governed completely by aristocratic selective principles.
5. The aristocratic selectivity in atomic valences follows the explanation given with my example from Japanese chess.
6. The standard expressed in Avogadro's law is something of a marvel. The length of the diameter differs from element to element. Under the same atmospheric pressure and temperature, the fact that no matter what molecule is generated by the elements, the process is standardized so as to produce an identical volume shows what we may think of as the magic of a super high-level principle of selection.

I will postpone a detailed description until later and would only note here that the kind of dialectics propounded by adherents of a crude materialism may be in for radical revision in the light of the progress that has been made in physics and chemistry.

Gamow's theory of the creation of the universe

Gamow writes that because a neutron cannot survive for more than twelve seconds on its own, there is no doubt that at the beginning when the material universe first took solid shape from light rays, all the atoms in the universe took shape within half an hour.[32] His view is backed up by theories that the structure of the atom was composed through a super-special level of selection with an a priori probability. This view been had partially expressed in Lecomte du Noüy's *Human Destiny*, which concurs from probability theory that atoms could not have come to a molecular structure by chance.

32. See his essay "Modern Cosmology," *Scientific American* 190 (March 1954): 54–63.

3 The Fitness of the Natural Environment

Henderson's theory of selective tendencies

Lawrence J. Henderson claimed that the natural environment is not blind, as Darwin had originally thought, but shows a tendency to a kind of strict selection, by means of which it possesses a degree of suitability or "fitness" that shows up even as a finality. On this point he stands no less firm than Wolfgang Ostwald of Berlin University.

Despite some lack of clarity in Henderson's philosophical thought, a consciousness of the evolution of chemical elements makes his insistence on positive selective tendencies aimed at producing a natural environment for the material world a rarity among biochemists. His influence is affirmed in the famous work by the Russian Vladimir Vernadsky, *Essays on Geochemistry*. The attempt to turn the recognition of tendencies into a teleology is one we find even in Julian Huxley.

Henderson sought to understand *tendency* as a third element along with the universe and life.[1] While he did not call his theory of selective tendencies a teleology and his approach was mechanistic throughout, he recognized that the theory implied teleology. Rather than *teleology*, he spoke of the suitability of the natural environment as *fitness*. The interesting thing about his insight is not his theory of teleological tendencies as such, but the way in which, from the standpoint of biochemistry, he was able to recognize that if the three elements of hydrogen, of carbon, and oxygen, along with their compounds water and carbon dioxide, form the basis for life, and if nature develops with life at the center such that its appearance constitutes the goal of the evolution of the cosmos, then the presence of tendency in nature is certain.[2]

---

1. See Lawrence J. Henderson, *The Fitness of the Environment: An Inquiry into the Biological Significance of the Properties of Matter* (New York: Macmillan, 1913), 280.
2. Ibid., 279–82; 280.

The fitness of the natural environment

Using Henderson's own words, we will present the summary of his theory found in *Fitness of the Environment*.

> II. Life is a mechanism (from the viewpoint of physical science). Accordingly it must be
>   a. Complex (physically, chemically, physiologically).
>   b. Durable, hence well regulated physico-chemically. This conclusion applies to –
>   1. The Organism.
>   2. The Environment.
>   c. Endowed with a metabolism. Hence there must be exchange with the environment of –
>   1. Matter.
>   2. Energy.
> III. The primary constituents of the natural environment are –
>   a. Water.
>   b. Carbonic acid.
> IV. In places where life is possible the primary constituents of the environment are necessarily and automatically formed in vast amounts by the cosmic process.
> V. Water, carbonic acid, and their constituent elements manifest great fitness for their biological role.[3]

The fascination of water and carbonic acid

> a. Water possesses a great number of unique or very unusual properties, e.g., thermal properties, solvent power, dielectric constant, surface tension, which together result in maximal fitness in certain respects, e.g., mobility, ubiquity, constancy of temperature and richness of the environment, richness of the organism in chemical constituents, variety of chemical processes, electrical phenomena, colloidal phenomena.
> b. Carbon dioxide possesses very unusual properties, e.g., magnitude of absorption coefficient, strength as acid, which together result in maximal fitness in certain respects, e.g., mobility, ubiquity, richness of the environment and organism in other elements and compounds, constancy of reaction, etc.
> c. Chemical compounds containing carbon, hydrogen, and oxygen possess unique properties, e.g., number, variety, com-

3. *Ibid.*, 208–9.

plexity, activity, variety of chemical relations and reactions, heats of reaction, instability, etc.

......

- g. The Evenness and Lack of Energy Change of the Process of Hydrolytic Cleavage,
- h. The Chemical Relationship of Carbonic Acid and Water to the Sugars,
- i. Instability of the Sugars,
- j. Variety and reaction of the Sugars.
- k. Heats of Reaction in Organic Chemistry,
- l. The Number and Variety of Compounds and Reactions of Oxygen with Other Elements,
- m. The Number and Variety of Compounds and reactions of Hydrogen with Other Elements,

[These] together result in maximal fitness in certain respects, e.g., as sources of matter and energy for the processes of metabolism, as sources of complex structures as the means of establishing complex functions, etc.

......

VI. Oceans are formed automatically in the cosmic process.

VII. The ocean possesses unique properties, e.g., mobility, richness in dissolved substances, durability, and stability of physico-chemical conditions, depending chiefly upon the properties of water and carbonic acid, which together result in maximal fitness in certain respects, e.g., as milieu, and as source of matter for the processes of metabolism, to moderate and equalize temperature, etc.

......

IX. There are no other compounds which share more than a small part of the qualities of fitness of water and carbonic acid; no other elements which share those of carbon, hydrogen, and oxygen.

X. None of the characteristics of these substances is known to be unfit, or seriously inferior to the same characteristic in any other substance.

XI. Therefore the fitness of the environment is both real and unique.

......

The fitness of the environment results from characteristics which constitute a series of maxima—unique or nearly unique properties of water, carbonic acid, the compounds of carbon, hydrogen, and oxygen and the ocean—so numerous, so varied, so nearly complete among all things which are concerned in the problem that together they form certainly the greatest possible fitness. No other environment consisting of primary constituents made up of other known elements, or lacking water and carbonic acid, could possess a like number of fit characteristics or such highly fit characteristics, or in any manner such great fitness to promote complexity, durability, and active metabolism in the organic mechanism which we call life.[4]

The theory of selective tendencies

Henderson concludes that "Out of the properties of universal matter and the characteristics of universal energy has arisen mechanism, as the expression of physicochemical activity and the instrument of physicochemical performance."[5] Note here his use of the term *arisen*. For Henderson, the fitness of the natural environment is an emergent process belonging to cosmic evolution. He goes on to speak in lofty tones of this fitness:

> If, then, cosmic evolution be pure mechanism and yet issue in fitness, why not organic evolution as well? Mechanism is enough in physical science, which no less than biological science appears to manifest *teleology*; it must therefore suffice in biology.[6]

As the passage makes clear, Henderson acknowledges purpose in the mechanism and tries to find a mechanism at work in purpose. Logically speaking, there are inadequacies in the explanation, not to mention the radical casualism he acknowledges when he counters:

> We simply cannot doubt that the origin of a body like the earth depends exclusively upon chance plus the properties of the elements, their relative amounts, the indestructible forces of nature, and other known factors of mechanism.[7]

The reason such a contradiction issues from the pen of one espousing a teleological position on the fitness of the natural environment is that he completely ignores the mechanical aspect of purpose. Since Darwin, biologists have endeavored to show their contempt for "teleology" and to consider "science" to be a mechanistic explanation of everything. But it is not possible to explain change in purely mechanical terms. Hence the

---

4. *Ibid.*, 267–73, 252.
5. *Ibid.*, 272–3.
6. *Ibid.*, 305; emphasis added.
7. *Ibid.*, 304.

Cosmic Purpose

introduction of that he calls "the tendency" in order to explain variation by means of blind casualism. Henderson has driven himself into this valley of contradiction in his attempt to "put the old teleology to death"[8] and rethink the framing of a teleology from a new theory of tendency:

> In short, our new teleology cannot have originated in or through mechanism, but it is a necessary and preestablished associate of mechanism. Matter and energy have an original property, assuredly not by chance, which organizes the universe in space and time.[9]

Claiming that the material earth appeared completely as a matter of chance, and at the same time that matter is not a product of chance, lands Henderson's "casualism" in a contradiction. And yet given the courage with which he has shown the fitness of nature, I do not wish to dwell too long on the inconsistencies in his logic. Let us attend rather to his new teleology: "Given the universe, life, and the tendency, mechanism is inductively proved sufficient to account for all phenomena."[10]

Evolution and the purpose in tendency

Adopting a theory of tendency similar to Huxley's, Henderson has tried to integrate it into a new teleology. In his theory of evolution Huxley was unable to refute the fact of teleological development. In the end, he had to acknowledge a tendency running throughout the material universe. When Henderson came to argue for a suitability in nature, even as he expressed satisfaction with a mechanistic explanation, he took a step away from those categories to come up with a new teleology in the form of a theory of tendency.

To be sure, evolution itself is a tendency. Above and beyond this, we have to acknowledge a certain slant in its orientation, variations in ascent and descent, efforts to preserve the tendency, corrective deflections made on its behalf, and the selective activities required to do so. Having already accepted a priori the work of selection in the natural world, and based on my earlier definition of "purpose," I am convinced that, even if teleology as such is rejected, its contents coincide with the demand for a scientific teleology.

Linguistic questions aside, in the absence of this content there is no

---

8. *Ibid.*, 311.
9. *Ibid.*, 308.
10. *Ibid.*, 308.

way to explain the natural world. Darwin thought that "nature is blind" and therefore explained evolution in terms of a blind natural process of sifting. Henderson, in contrast, has shown us that nature is *not* blind, that the universe is purposive and pursues a strict course of tendency to give birth to life.[11] He did not speak as such of "purpose in tendency," but given the new theory of teleology he proposed, we may indeed refer to it as a "tendency theory." To be sure, his contribution to the world of biochemistry is enormous, and by demonstrating a tendency involved in selection—or as he called it, "fitness"—at the organic and inorganic levels through the three elements of hydrogen, carbon, and oxygen, his impact on geochemistry and general chemistry has been considerable.

Tendency in water

The water of life

The cosmic finality of water appears in the contributions it makes to life. The smallest unit of life is the cell, and water, more than anything else, is the key ingredient in cellular structure.

Life has a three-way relationship to gases, liquids, and solids, but when it comes to the cells that make up the body, the fact that as a liquid water dissolves colloids has to be considered its principal trait. The idiom "water to a dying person's lips" points to the fact that at the moment of breathing one's last, nothing is more desirable than water.

Among the seven sages of ancient Greece, it was Thales, the originator of trigonometry, who said that "All things are born of water and return to water." From of old, water has been given a special place.

If we think of water only as a chemical compound of hydrogen and oxygen, we face a gaping abyss. Oxygen boils at a temperature of -181°. Hydrogen boils at the still faster temperature of -253°. From this standard, we would expect water to freeze at around -150° and to boil at -100°. But when oxygen and hydrogen are joined together, an extraordinary change takes place to sap its strength: water has the quality of freezing at 0° and boiling at 100° and above.

Scientists like David Burns have noted that a water molecule, H_2O, has the unusual trait of changing according to the temperature and the material it dissolves. That is to say, when the temperature is lowered, its

---

11. *Ibid.*, 312.

molecules are thought to increase in size, and when it rises, to be simplified. Logically speaking, we have the gas (H_2O) and the solid (H_2O)^3, but as the temperature rises, there is a further breakdown from (H_2O)^3 to (H_2O)^2 which is dissolved and evaporated into H_2O. In many cases involving elements of a less extreme character than water, the result is a highest or lowest valence. By and large, the highest valences show up as ordinary liquids and carry out the solvency function of salts.

At first glance, the existence of water, the element with the deepest connection to life, seems to be a matter of chance. A closer look shows something surprising in virtue of the following factors most necessary for life: (1) electrical energy, (2) ionization, (3) the occurrence of the hydrogen ion, (4) the cooling apparatus for radiation, (5) the capacity for colloidal dissolution, (6) the special catalytic character of water, (7) the permeation of cellular membranes, (8) heat insulation, (9) expansion, (10) cyclicity, and so forth. The mission and finality of water is manifest in its capacity for selection and regulatory buffering action, and in laying the groundwork and opening up the most suitable environment for the appearance of life.

Water is composed of 2 parts hydrogen and 1 part oxygen. (We omit a discussion of deuterium or heavy hydrogen here.) The method of their compounding is more mysterious than we could have imagined. In inorganic chemistry, when we write H_2O, we usually think in terms of a compound that puts hydrogen at the sides and oxygen at the center, with all of them on the same level. Following Pauling's rules of chemical

Dielectric constants[12]

air		1.0	water	81.7
carbon dioxyde gas	CO_2	1.0004	alcohol	25.0
hydrogen		.9997	formaldehyde	84.0
			acetic acid	6.46
			HCN (liquid)	95
			vaseline	
			turpentine	2.2
			benzine	
			paraffin	
			liquid fat (animal)	3.0–3.2

12. See David Burns, *An Introduction to Biophysics* (London: Churchill, 1921), 53.

Relative speeds for hydrogen and sodium hydroxide ions[13]

Negative absorption (cations) +			Positive absorption (anions) −		
Ion	Atomic weight	Relative speed	Ion	Atomic weight	Relative speed
H	1	318	OH	17	174
K	39	64.6	$\frac{1}{2}SO_4$	48	68
NH_4	18	64.4	Br	80	67
$\frac{1}{2}Ba$	68.5	55	I	127	66.5
$\frac{1}{2}Sr$	43.7	51	CL	35.5	65.5
$\frac{1}{2}Ca$	20	51	$\frac{1}{2}C_2O_4$	44	63
$\frac{1}{2}Mg$	12	45	CH_3C_{oo}	59	33.7
Na	23	43.5			
La	7	33.4			

compounding, oxygen is at the center and the two hydrogen atoms at a 90° angle to each other. Even though both hydrogen and oxygen are combustible, when the atoms are combined, they miraculously convert to a noncombustible "radical."

Besides this, water's greatest contribution to life must be said to lie in its electricity, as direct experiment confirms. Rather than say that water itself possesses electricity, we may perhaps better speak of its aiding in the production of electricity through ionization and solvency, as in the production of electrical energy like a miniature light turning on when water is mixed with a small amount of common salt.

Pure water on its own, of course, does not have the power to produce electricity. But water shows a surprising polarization, which is said to "correspond to the square of the refractive index of a body." If we take the refraction of air as an index and set it at 1, the refractive index of water it is 81.7 times greater, or nearly 9 times that of air. Some liquids like formaldehyde and hydrated liquid cyanide show an index slightly higher than water, but among ordinary liquids, nowhere is the index as high as it is in water.

13. *Ibid.*, 50.

Furthermore, within liquid solutions the velocity of water ions is far higher than that of ions in other substances. Compared with kalium (K) at 64.6 and natrium (Na) at 43.5, the velocity of substances with hydrogen ions is an astounding 318. In addition, when the hydrogen ion is released from water (OH), the velocity of its adhesion to a positive charge is extremely fast—174 in comparison with 66.5 for iodine and 65.5 for chlorine.

Here again it is easy to understand how water strives to contribute to life through its electrical qualities.

Solvency and ionic quality

This electrical solvency and ionic quality of water is the reason all living things have a relationship to water. Plants and animals cannot appear without water. Even in the driest of deserts, as long as there are a few drops of water, luxuriant plants will always show up. Or again, research into the evolution of the sexuality that determines male and female in the plant world shows that the earliest plants were produced in water and only later evolved to life on dry land.[14]

The late Nagai Hisomu, professor of physiology at Tokyo University, wrote in his book *A Scientific View of Life* that "sea water is roughly thirty-seven percent ordinary salt, in accord with whose absolute concentration, osmotic pressure is by far higher than that of the interstitial fluids of high-order life forms; and in not a few cases the proportions of 'salt' ions is the same in both of them...." The following chart by E. V. McCollum purports to show these relationships.

McCollum's table.

	Na	K	Ca	Mg
sea water	100	3.6	3.9	12.1
jellyfish body	100	5.2	4.1	11.1
fish serum	100	3.7	4.9	1.7
canine serum	100	6.9	2.5	0.8

In lower animal forms like the jellyfish, the absolute concentration of salt as well as its relative concentrations conform completely to those of salt water. The concentration of salt in the blood of higher animals is about 12 percent or about one-third of what it is in sea water. Even so, if

---

14. See H. G. Wells, ed., *The Science of Life* (London: Waverly, 1929–1930), 3 vols.

the weight of natrium (Na) is set at 100, the relative weight of kalium (K) and calcium (Ca) by and large coincides with that of salt water. These relationships are roughly equivalent to those found in a Ringer's solution.

Or again, if we assign natrium a value of 100 and compare it to the K and Ca ions, salt water, the blood of higher animals, and the Ringer's solution all show a value of 2 for K and 2 for Ca. Furthermore, if we dilute salt water and adjust its osmotic pressure to that of blood, it serves as an ideal physiological saline solution. Only the concentration of magnesium (Mg) is much higher in salt water than in the interstitial fluids of the higher animals. These facts are far too important to close one's eyes to as if they were a matter of chance.[15]

Gustav Von Bunge and E. V. McCollum compared earth today with the earliest Cambrian Period when the continents were formed of earthen crust, drawing attention to the increased amounts of salt and magnesium:

> The salt in the soil is continually flowing back into the oceans through rivers and streams, so that the salt in the oceans is continually increasing. But as water evaporates and turns to rain, the absolute volume of water in the oceans remains more or less fixed. For this reason, the sea water inhabited by the ancestors of the animals of the Archeozoic Period that were born in the sea—namely, the blood of present-day animals—obviously differed from the sea water of today and the absolute volume of salt contained in it. The relative scarcity of K in the oceans is due to the fact that the time after the Cambrian Period was marked by a flourishing of vegetation on dry land, which meant that large amounts of K were absorbed by plant life while the amount of K that flowed into the oceans was correspondingly reduced. As for Mg, it was not so much that animals came onto the scene as it was the subsequent and relatively quick apparent increase of Mg in the oceans; even at present it may be said that the increase of Mg in the sea water is large in comparison to what we see in other elements.[16]

In the light of this, it is easy to understand the important role played by water. That being the case, why and in what manner did water become so crucial to life? There are ten reasons and more for this, but I believe its greatest task consisted in its adjusting the alkalinity in cells by isolating the water ion H and the water acid ion OH.

---

15. Nagai Hisomu 永井 潜,『科学的生命観』[A scientific view of life] (Tokyo: Shunjūsha, 1929), 165–6.

16. See Shōji Rinnosuke 正路倫之助 and Yoshimura Hisato 吉村寿人,『生物の物理化学』[The physical chemistry of biology] (Tokyo: Nippon Hyōronsha, 1931), 349.

[OH'] sodium hydroxide ion	[H] sodium hydroxide ion	pH	
10^{0}	10^{-14}	14	strong ↑ alkali
10^{-1}	10^{-13}	13	
10^{-2}	10^{-12}	12	
10^{-3}	10^{-11}	11	
10^{-4}	10^{-10}	10	
10^{-5}	10^{-9}	9	↓
10^{-6}	10^{-8}	8	weak
10^{-7}	10^{-7}	7	neutral
10^{-8}	10^{-6}	6	weak
10^{-9}	10^{-5}	5	↑
10^{-10}	10^{-4}	4	
10^{-11}	10^{-3}	3	acid
10^{-12}	10^{-2}	2	
10^{-13}	10^{-1}	1	↓
10^{-14}	10^{0}	0	strong

Although warm-blooded animals met with a drop of temperature from around 35 to 41 degrees, it is still possible to tie them to life. However, the 7.3 to 7.5 density of the pH (hydrogen ion) neutralizing negative ions and positive ions can truly be called electrical, and with only a modicum of difference, it allows for the continuation of life.

Taking this into account, marked variations in the struggle for survival, fluctuations in the atmosphere, changes in the food supply, and accommodation to the conditions for life still leave considerable work to be done. The density of the hydrogen ion, for example, helps us understand the seriousness of environmental adaptation. That is to say, in order to bring the density of the hydrogen ion to a level of 7.3 to 7.5, the addition of phosphorus to the interior of the cell, the use of bicarbonates or proteins, and the free use of iron-enriched chromoproteins show the hard work required to enable adjustment and balance.

When I consider the marvelous success of all this and the gravity of the adjustments that had to be made, I am unable to concur with the likes of Darwin and Huxley that natural selection has taken place as a matter of chance. Quite to the contrary, I agree with Lawrence Henderson that we have to think in terms of the "fitness of the environment."

I believe these adjustments are absolutely inexplicable without giving consideration to a finality aimed at life.

The mystique of water is by no means the end of the story. As George Crile notes in his splendid book *The Phenomenon of Life*,[17] the nissl bodies (iron colloids with oxygen) contained in brain cells form by compounding with nitrogen iodide sent from the thyroid and through electrons released by ultrashort-waves. These ultrashort-waves determine direction and collide with the white matter of the brain. One need only think of the memories stored there. But because the electrons so released have a high temperature, the mission of cooling this high temperature is carried out by the water contained in the brain. Crile argues that the cerebral matter of young children is about 85 percent water, and that the greater the amount of water, the more active children can be.

The hydrophilic action of matter

Water has a wondrous capacity to dissolve colloids. Naturally, there is also an aversion to water that is diametrically opposed to the hydrophilic quality of colloids, that is, to their habitual affinity for water. The hydrophilic quality is often found as a water mantle in invertebrate membranes.

The water present in cells serves as a clear display of its colloidal nature with respect to salts. By increasing the *surface* of a colloid, it is possible to augment its electrical charge. Even in small cells, several million colloidal particles are dissolved in the effort to contribute to the capacity for life.

Without detailing the activity of colloids dissolved in water, we may learn from colloidal chemistry how colloids precipitate and lose their electrical charge when alcohol penetrates a cell. Moreover, it is reported that by permeating the cell bacteria producing trachoma, liquid solutions of zinc sulphate and boric acid cause the colloids to precipitate and have the power to make them collapse. Thus we know that only water works the marvelous effect of releasing colloids and has the power to ionize.

The catalytic activity of water

We should also note that water shows the same kind of catalytic activity as platinum black and iron. This is known as hydrolytic action. Nor must

---

17. George Washington Crile, *The Phenomena of Life: A Radio-Electric Interpretation* (London: William Heinemann, 1936).

we forget that the most necessary digestive activity of living organisms depends on the functioning of various sorts of acids, an activity that is always accompanied by hydrolysis and makes use of water as a kind of catalytic agent. In the process, water never loses its properties as water. On the contrary, the truly amazing thing is that were water to lose its special characteristics, digestion would not be possible.

In most cases, rain contains negative electricity, even if only in very small amounts. In rain water the amount of negative electricity often increases. During a stay in India, George Simpson, director of the London Meteorological Observatory, measured the electricity carried in the almost daily assault of heat thunderstorms in Simla, the capital city of Himachal Pradesh in northern India. His results show that during the hours of rainfall, 71 percent of the electricity—or 75 percent of the volume of electricity—carried a positive charge. In other words, much of the rain in thunderstorms is electrically charged.

At the same time, Simpson examined in detail the phenomenon of electrical detachment through the separation of raindrops, which Philipp Lenard had discovered in his laboratory. At first he found that when raindrops without an electrical charge were separated, negative ions of -0.0033 units of static electricity and positive ions of +0.0011 units of static electricity resulted. The accompanying differential of -0.0022 units of positive electrical charge—namely, the positive electrical charge of +0.0022 units of static electricity—occurred in raindrops produced at the time of separation.

The separation of raindrops thrives with a rise in electrical current of 8 millimeters a second and above, and in the smaller raindrops that have separated out, electricity with a positive charge increases in positivity even as a negative charge gradually emerges, making it clear that the rise in electrical current in cumulonimbus clouds occasions thunder.[18]

Lenard also measured the phenomenon of negatively charged electricity occurring in the surrounding air, when the drops of water we find in waterfalls and the like fall at a rate of eight meters a second, producing electricity through the separation of droplets. Lenard drew attention to the way in which this "waterfall effect" is based on two strata of electricity, H and O, in the separation of electrical charges.[19]

---

18. See Nakaya Ukichirō 中谷宇吉郎, 『雷』 [Thunder] (Tokyo: Iwanami Shoten, 1950), 177–8.

19. Ibid., 172.

Finality in Rocks

Vernadsky's theory of rocks

V. I. Vernadsky has written what we may call the "bible of rocks." At one time professor of Mineralogy at Moscow University, Vernadsky was exiled from Russia during the revolution of 1917 and took up a post at the Sorbonne in Paris where he lectured on geochemistry.

His famous work, entitled *Geochemistry*, has become a veritable "rock gospel" in the field. Vernadsky approached the cosmos as a whole, not from the heights but by making quantitative calculations of the rocks that form the earth's crust. Measuring the depths at which rocks were assimilated into life itself, he was led to conclude that the universe "is not the result of chance but a product of the order that governs nature."[20]

He began by identifying six classes of elements that go into the makeup of rocks: (1) precious gases, (2) precious metals, (3) environmental elements, (4) dispersion elements, (5) strong radiation elements, and (6) rare earth elements. He further drew attention to the fact that among these environmental elements, which account for about half of the elements, all have even atomic numbers, among which the most active have an atomic number divisible by 4.[21]

Furthermore, Vernadsky observed that 99.7 percent of surface rocks are made up of environmental elements, while elements related to dispersion account for a mere 3 percent of their total makeup. In addition, the forty varieties of environmental elements are concentrated in living organisms, and after the death of an organism, some of these survive in the form of surface rocks. In particular, he noted that magnesium veins and deposits of certain types of iron, limestone, and vanadium phosphate are all transformations of the corpses of microorganisms. This led him to the claim that "the total sum of surface rocks and living organisms are always equal."[22]

The constancy of environmental elements

Vernadsky subscribed to the popular nineteenth-century belief in the cooling of the earth's crust. He also held to the scientific view that the

---

20. Translated from the Russian into Japanese by Takahashi Jun'ichi 高橋純一,『地球科学』[Geochemistry] (Tokyo: Uchida Rōkakuho, 1933), 444.
21. *Ibid.*, 29, 33.
22. *Ibid.*, 223.

environmental elements that composed the earth's crust preserved a numerical constancy from the time the planet was formed.[23] In other words, he was convinced that not only the oxygen, silicon, iron, aluminum, and so forth that account for about 70 percent of the earth's crust, but also sea water, carbon, and indeed all the other forty-eight or so environmental elements maintain a constant quantity.

As a result, he theorized that material substances constructed from the rocks that are formed of these elements also show a quantity that has not changed from primitive ages down to the present.[24]

Vernadsky made a momentous contribution to geochemistry in his discovery of the so-called "kaolin core" that is formed by an anhydride made up of silicon and aluminum and has an unchangeable quality. He proved that kaolin core with a molecular formula $AL_2Si_2O_7$ underwent a mutation to develop into the silico-aluminum anhydride $AL_2Si_{2+2n}O_{7+2n}$. He further pointed out that the feldspar that accounts for about 60 percent of the earth's crust has a molecular formula of $AL_2Si_2O_{15}$. In addition, he believed that in the absence of a disintegrating biotic force, ring-forming molecules cannot easily be destroyed. He depicts kaolin core as follows:

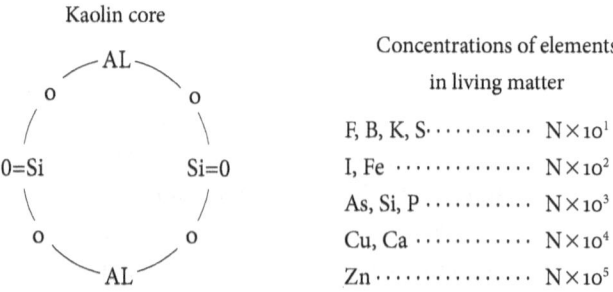

The heat-absorbing function of the circular kaolin core had a momentous impact on the history of the earth. The principal factor in the achievement of the thermodynamic stability that the earth enjoys today is thought to lie largely in the development of kaolin core by this stratum of rock composed of silicon and aluminum. Because what Hans Suess called the "Si-AL stratum" shifted upwards from the relatively heavy part of the earth's core with its heat-absorbing character, it was able to pre-

---

23. Ibid., 442.
24. Ibid., 313.

serve a thermal balance above red-hot lava and offer a place habitable by living organisms.

This amazing structure becomes all the more interesting when we consider the relationship between the physical body needed for a living organism and the nature of silicon and aluminum. Vernadsky noted that rocks composed of silicon and aluminum as well as colloids were absolutely incapable of compounding with water, and Suess confirmed that the total volume of water on the earth remains constant and that the volume of sea water is also fixed.

But Suess did not believe, as other scientists had, that sea water and primitive animals were made of the same stuff. He consistently upheld the Redi principle that "life comes from life" and that the concentration of atoms in living matter is radically different from the degree of atomic dispersal found in sea water.[25] As shown in the table above, he did not recognize concentrations of Na, CL, and Mg.

Inasmuch as he found the material in silicon and aluminum enormous enough to absorb carbon dioxide, Vernadsky thought that soil breathed. Given the ability of both salt water and fresh water to dissolve carbon dioxide, he noted that nature completely regulates gases like carbon dioxide necessary for vegetation. He further asserted that the regulation of nature was equivalent to the regulation that occurred through living organisms.[26] In particular, silicon purifies carbon dioxide gas to a surprising degree, and the kaolin clay found one hundred meters under the surface of the earth absorbs four and a half times the amount of carbon dioxide in the atmosphere, while conversely, through volcanic eruptions, carbon dioxide is sent back into the atmosphere. He showed how about 97 percent of the gas erupting from inactive volcanos is carbon dioxide and that roughly the same percentage holds for active volcanos.

Given that rocks composed of silicon and aluminum are water repellent and show a selective affinity to carbon dioxide, Vernadsky could not deny a kind of finality present in this mission to develop habitable land for life.[27]

---

25. *Ibid.*, 322. [The Redi principle, named after the seventeenth-century physician Francisco Redi, is treated on pages 355–8.]

26. *Ibid.*, 253; cf. 334.

27. *Ibid.*, 138, note.

The silicon cycle and the history of rocks

The water, carbon dioxide, and silicon rock needed by living organisms noted above led Vernadsky to describe a cycle in the earth's crust:

> Things living on the earth's surface continually absorb aqueous solutions in a fluid state, that is, silicon atoms in a state of colloidal suspension or *Sol*. It has been confirmed that the protoplasm of living cells even absorb minute particles of quartz and opal in small amounts.
>
> Certain types of living organisms function to destroy the stable kaolin core of solid clay and to absorb silicon atoms. Furthermore through the activity of anhydride carbonic acid and water, as well as the vital activities of the organisms themselves, silicon is concentrated into a colloidal state, becomes opal, or crystallizes into quartz, or else it takes on a fluid form in water and further produces free acids and the like possessed of colloidal soil, hydrous silicic acid magnesia, clay, and kaolin core. Through biochemical action, hydrous silicic acid magnesia also absorbs soil separated and dispersed from silicic acid by the destruction of kaolin core.
>
> In the passage of geological ages, these silicic acid elements began to undergo a metamorphosis. Hydrous silicic acid magnesia became serpentine, talc, and chlorites, finally converting to augite and amphiboles. On the other hand, kaoline core clay returned to its primitive state and gave birth to mica and feldspar. Opal was changed into crystal quartz. In the depths of the metamorphosing stratum, kaolin core alumina silicon minerals synthesized through the activity of aqueous solutions under a high pressure of elevated temperature. This all took place on a large scale within magma which brought about melting. In this way, by borrowing the energy of the sun and the power of magna, the cycle of silicon, which makes up the main part of the earth's crust, continues on forever.[28]

Rocks are alive

Readers of Vernadsky's *Geochemistry* are struck by the degree of vitality in rocks as witnessed in their cyclical quality and their functions as stimulative agents. From the time he served as director of the Leningrad Institute for Biochemistry, he was clearly knowledgeable about the relationship between the rocks of the earth's crust and living organisms.

Referring to reports of the German zooligist J. V. Carus, that the volume of orthopetera flying over the Red Sea from Africa into Arabia was enormous, reaching 4.4×10^7 kilograms and covering a land surface of

---

28. Ibid., 193–4.

5.97 square kilometers, Vernadsky recorded that in the hundred years that had elapsed since the middle of the nineteenth century, the total amount of copper, iron, and zinc mined and refined came to 4.4×10^7 kilograms.[29]

He further drew attention to the vigor and vitality of microorganisms, as well as to the fact that diatom, unimpeded, would cover a land surface equivalent to that of the entire earth, and that in five days one type of infusorian paramecium was able to cover an area 10^5 times the surface of the earth.

He also pointed out that living organic material is a good regulator of changes in the earth's crust, never forgetting that nature itself shows roughly the same regulatory capacity as the life force. This prompted him to follow Henderson by attempting to explain geochemistry from the standpoint of a new teleology.

A new teleology

Vernadsky summarized Henderson's ideas in these words:

> There is a view that has become conventional in the world of biology today following the lead of the American physiologist Henderson, who has recently expounded his interesting ideas clearly and in detail. Not only does Henderson point out the intimate relationship of water to living organisms, but he goes on to demonstrate that water holds a unique place among the multitude of compounds. Indeed, in virtue of its physical and chemical qualities, water has to be seen as an extremely important and characteristic ingredient in the survival of organisms.
>
> Of all the chemical ingredients, nothing can stand up to water when it comes to the relationship to living things. Water alone has its destiny bound to living things in virtue of its inherent qualities.

This idea may be seen as a revival in new form of the old teleological ideas circulating in the middle of the nineteenth century, according to which there is a general harmony in the whole of the natural world and the order among everything in the universe is not a matter of chance. At this point, Vernadsky paraphrases Henderson:

> As contemporary science shows, living organisms inhabit an environment constructed for them on earth which has water as its most important element and to which they are perfectly suited. Water itself arose in the course of the evolution of the universe as something

---

29. *Ibid.*, 61.

suited to all living phenomena. Moreover, its fitness is of such significance that no other fitness gleaned from the evolutionary process can compare to it.

These remarks constitute less a formal teleology than a theory based on experiments and on important facts related to the study and understanding of life. The earth's crust is not merely a complex agglomerate of materials. Complex it is, but it has also been certified to possess a well-regulated selective function:

> The activity of living phenomena has a close relationship with the earth's crust whose foundations may be sought not only in the phenomenon of the prosperity of living things over many years, but also in the phenomenon of the geological development of the earth's crust. In recent years these foundations have come to be sought in the atomic makeup of the earth's crust.[30]

Thus Vernadsky's work in geochemistry has shown us the way in which the rocks that formed the earth have an intimate connection to life by offering a new way to view the finality of rocks in relationship to life.

A note on the benefits of volcanos

The volume of nitrogen on earth, equivalent to one square mile, is forty times the amount found on the entire planet of Mars. Carbon dioxide, another most important nutrient for plants, is brought to the surface for them by volcanos. Were the supply of carbon dioxide to vegetation on our earth to be cut off, the plant world would become extinct, which in turn would result in the indiscriminate extinction of human and animal life. Mars is older than planet earth and clearly there is almost no volcanic activity there today. Here we see how very important carbon dioxide is.[31]

THE FITNESS OF SOIL FOR LIFE-FORMS

Soil converted to rock in a brief period

One is amazed to learn that in India the entire portion of silicon in red soil is flushed out with the rain, leaving only minute particles of aluminum and iron. If you dig about 3 feet into the ground, shovel out from there on down whatever amount of red dirt you want, and then let

---

30. *Ibid.*, 41–2.
31. See E. C. Pickering, *The Youth's Companion* 1/6 (1913): 10–11.

it dry several hours in the tropical sun, you end up with a splendid material for construction. In southern India this is known as *laterite* (from the Latin *later*, meaning brick).

In the poor villages of southern India, houses and Christian churches are constructed of this laterite. A church can be constructed for as little as 50 yen at the pre-war rate.[32]

Such is the curious quality of soil that the heat of the sun and weathering combine to cause corrosion and produce colloid, which in India produces rocks with the flexibility of a rubber eraser or clay. Where magnetism works in virtue of the electricity present among minute particles of soil, this malleability is lacking. This is easy to understand from the colloidal quality of soil. Physiology teaches us that colloids are also produced in the human body. The book of Genesis records that humans were made out of earth, and colloidal chemistry confirms it.

Colloidal chemistry also speaks of the "acidity" and "alkalinity" of soil. Things with a positive electrical charge are referred to as acidic and those with a minus charge, as alkaline.

White soil in frigid zones

The soil found in the tundra of the two polar regions is white. Tundra also formed in the Ishigari plains of Hokkaidō. The tundra is comprised of peat moss, which can be used for soundproofing. In order to convert these regions to farmland, either topsoil has to be imported or dairy cows have to be pastured and made to feed on the peat moss. The stomach contents released in their excrement contain disintegrating bacteria which is necessary to convert the peat into soil. Without such methods the tundras of Manchuria cannot be converted to arable land either. The way land is prepared for cultivation in warmer regions does not apply to white-soil regions, where pasturing cows and waiting for several decades produces a top soil well suited to cultivating rye and oats.

Between white soil and red soil

Between the white soil in the polar regions and the red soil found in areas around the equator lie the vast desert regions of chernozem and podzol.

Many things found in the desert regions, depending on the color of the base rock, have a largely grayish color. Everything I saw in the

---

32. [The equivalency in US dollars at the time would have been less than $220.]

Cosmic Purpose

Mongolian desert as well as in the deserts around India, Asia Minor, and Egypt had this gray color. The desert regions of Australia are also gray. Yet I was surprised to find powder-like red laterite soil on the plains of Paraná in South America. Travellers through the state of Paraná in Brazil have to be resigned to having their shirts and trousers dyed red.

As all scientists tell us, as long as there is water, desert land is suited for farming. "Geological water" runs beneath the surface of the desert. The water under the Australian desert is rich in salt and can be extracted for use. Taking advantage of new chemistry, the salt can be removed to make the water potable. As long as water is present, organisms can thrive.

I had occasion to visit the volcanic region in the south-central desert of Australia and felt as if I had stepped on to the moon. I was surprised to see that in the absence of water, the bare ground was eroded, giving it a structure identical in form to lunar volcanos. I thought it must be like the way things looked at the beginning. There is a volcano in the Arizona desert near the Albuquerque airport that was formed by a meteor that had fallen from the same space as those that had fallen into the lunar world. Pieces of meteor rock are sold at the airport for souvenirs.

In the desert of Utah water can be drawn to the surface from several hundred meters underground to feed tens of thousands of hectares. It is a treasure for the Mormons who work tirelessly there. The melting snow of the nearby Rocky Mountains is carried in strata and flows underground where it can be beneficial for farming, which is why underground water needs to be given attention in soil research.

History shows us that the Mongolian desert, along with the deserts of India and Paraná are "man-made deserts." The large canal of several hundred miles dug in the barren lands of North America to carry the waters of the Colorado River and create beautiful farmland is altogether an enviable achievement. India and Egypt tried to imitate the feat. As the visitor to India will see, prior to independence the British government tried to carry water to the desert lands for use in irrigation.

I believe that if the methods of bringing underground water to desert regions were tried in China and Mongolia, it might also be possible to produce arable land there. If the vast sums of money spent on warfare were diverted to such projects, it would be easy to open land suitable for human habitation.

If we exclude the white soil of the tundras, the red soil of the tropics, and ashen desert soil, only the podzol and chernozem regions are left.

The Fitness of the Natural Environment

The podzol region

To the south of the white-soil regions, in the forests of Siberia where the Russians put Japanese and German prisoners to work as lumberers, lies the podzol region. Not having seen the podzol regions in the southern hemisphere myself, I cannot vouch for conditions there, but my impression from travels through Sakhalin, Finland, Norway, and Sweden is that if indiscriminate logging is not stopped, the consequences for the human race will be calamitous.

The great forests of Sakhalin were largely turned to desert by the Japanese. The same fate is said to await Siberia. The soil of these regions is strong in acid, shows little soil erosion, and is pale brown in color. But if lime is scattered to neutralize the acidity and if legumes are planted to feed the roots with the air-borne azotobacter needed to grow grass, it becomes splendid farmland.

It seems to me that the cultivation of plants in the legume family must by no means be neglected. For the cold regions, these include first of all acacia. By planting the acacia in three or four rows, not only can they serve as a snowbreak, but by cultivating filbert around the roots, they can serve as fodder and even be harvested for human consumption. Since these woods can also serve as a windbreak and snowbreak, it would seem necessary to cultivate legumes and filbert, not just conifers.

In planting conifers, one must plan ahead against difficulties, including against long years of starvation, by cultivating trees like the edible pine nuts of the Rocky Mountain pinyon and the pine trees of India and Nepal, or the edible pines of Korea's Mount Kumgang. Peoples like the Japanese who have long lived in warm regions are ill-informed of farming methods in cold regions and hence are ignorant when it comes to using the podzol region.

The chernozem region

To the south of the podzol region an area has developed that may be called "the world's granary." In Eurasia and America such areas are called "black regions." Grasses spring up in the humidity. On the North American continent, a brown region lies to the south of the black region. As in the state of Iowa, corn grows as high as one's head.

Going further west, there once lay a great bay in the region between the Rocky Mountains and the Sierra Nevada where the waters of the Pacific Ocean had penetrated during the Paleocene period. A vast plain of limestone formed through the rising of the earth's surface, comparable to the Eurasian continent with its stratum of limestone in the Alps and to the desert region of limestone that grew up in Mesopotamia, Iran, and Iraq, in the lower part of which the major oil fields of the earth were formed.

The beneficial mud of the Ariake Sea

The crucial ingredients in the formation of soil are (1) silicon, (2) aluminum, (3) oxygen, (4) hydrogen, (5) iron, (6) calcium, (7) magnesium, and (8) potassium. In addition, there are things which, while not important to the constitution of plants, nonetheless are indispensable as micromolecules. Although usually overlooked, there are six such elements: (1) iron, (2) zinc, (3) copper, (4) boron, (5) manganese, and (6) molybdenum. Nitrogen, phosphorous, potassium, calcium, magnesium, and sulfur need to be present, but they are always in short supply in ordinary soil and continually need to be supplemented from without. This is the benefit of using fertilizer.

Of late, nutritionists have stressed that the consumption of plants lacking in these six kinds of microelements causes sickness in animals and people. Traditionally, when Japanese farmers finished planting, they would go to a hot springs and take a rest for several days. This old custom may be grounded in the need to imbibe microelements dissolved in different fluids.

Besides the six elements enumerated above, iodine counts among the microelements we humans need. Without iodine, Graves' disease sets in. I was surprised to learn that roughly 10 percent of young girls living in the Australian desert suffer from the illness. I once instructed young patients in Japan with Graves' disease that "if you drink sea water for a week, you will be cured." I am informed that those who followed my advice found that the swelling in their necks subsided and that the disease indeed cleared up.

In the Kyūshū University Hot Springs Research Center located in the city of Kyūshū Beppu, researchers have smeared a cloth moistened with mud from the Ariake Sea on the bodies of patients, with the amazing result that it cured their neuralgia. It is also possible that the microelements dissolved in the mud had been absorbed into the skin. If diseases can be cured from the outside, the intake of food would seem important to health. Soil is living. It is a colloid, a treasure house containing every kind of protistan and air-borne azotobacter. Soil absorbs carbon dioxide and neutralizes poisons.

Soil arose on the surface of the earth for plants to grow in. The idea of a plan so devised that the people who ate those plants were able to inhabit the earth points to a heavenly wisdom a bit beyond the reach of ordinary common sense and leaves me with a profound sense of admiration.

Like the people smeared with the mud of the Ariake Sea, farmland and rice fields need to be smeared with mud. I thought of this on hearing that harvests in the fertile fields to the north of Lake Biwa in Japan's Shiga Prefecture amount to a mere 2 *koku* (12.74 cubic feet). This is due entirely to the inadequate amount of microelements. If microelements could be added to crops by carrying in soil from far away mountains or compost made from mountain vegetation mixed into the mud, the harvest might be increased many times over. It is the same for rice and barley as it is for human beings. If the invigorating effect of microelements is insufficient, a disease known as rice blight sets in. Unless the mud is prepared, the crops and the people who consume them all get sick.

Latent Purpose in
the Structure of Life

4 Latent Purpose in the Structure of Organisms

ADAPTATION IN THE INNER ENVIRONMENT OF ORGANISMS

The adaptability of the structure of life

That things possessed of life should have appeared in a random universe is a marvel. I have studied the reasons for the arrival of organisms on earth by looking at adaptation in the natural environment. In doing so, it was necessary to examine the situation that led to the birth of living things able to stand on their own, independent of the natural environment.

No matter how the elements of chance are arranged, they remain chance. Standing independent of a natural environment thought to have come about by chance, living entities that involve metabolism, physical development, growth, reproduction, and multiplication each have distinct physicochemical and mechanical features with their own structure. The living body itself is a *microcosm*. Not without good reason did the German philosopher Hermann Lotze publish his research on "human beings" a century ago under the title *Mikrokosmos*.

How could living things supported by the natural environment but not identical with it have appeared? Such things live in the world of nature and are governed by the laws of physical chemistry—and yet they transcend it. Besides these biological laws, laws of psychology are also at work here. Call them biological or psychological as you will, these laws are extremely complicated, concentrated, and mechanized by the laws of selection. The point at which the inexplicable governance of choices comes into the picture is the point at which we have to think in terms of *purpose*. Of course, for the sake of convenience these features are expressed mechanically and in terms of physics. Nevertheless, we need to acknowledge that the mechanism that organizes selectivity is itself purposive.

The origins of life

Explanations of the origins of life are many and varied. Arrhenius surmises that "it probably came from outside of the solar system." Others think that it was constituted in a single and never to be repeated geological event. The illustrious French chemist Louis Pasteur thought that "life is born of life and not of inorganic material." Oparin, in contrast, argues that life emerged from within the natural world. Although I myself find nothing to object to in the idea that life was born from within the natural world, this entails considerations that differ completely from the way this claim has been understood up until now by only considering materialist explanations.

There are serious difficulties with the view that life came about from simple matter. If we were to grant that the phenomenon of life represents a unified focal point at which the contributions of matter in the natural world and all the laws of chemistry come together, there would be no need to look for reinforcements outside of the world of nature. And yet there is no doubt that combining all the physicochemical laws in the cosmos and concentrating them on a single point is more than a simple material event. In fact, the unifying force was brought about through a priori selection. It is also a fact that such a priori selectivity transcends nature. All laws, precisely because they are not human creations, represent a supernatural phenomenon.

The analysis of cellular physiology

To take only the biology of a single cell, if all the physicochemical laws of solids, liquids, gases, electricity, magnetism, and radiation are unified and made functional there, as I believe they are, then it is unreasonable to think of life only in terms of materialistic theory. I accept that energy was transformed into matter, but I also accept that laws permeating matter are at work. It is not only that the various laws of physics and chemistry are active throughout the three states of matter (solid, liquid, and gas). We must also include the fundamental influence on "living bodies" of colloidal electromagnetic energy and also radiation that shapes these three states:

> *solids*—colloidal crystal gel, proteins, microelements
> *liquids*—liquid crystals, quasi-liquid crystals, water, *Sol*
> *gases*—hydrogen, oxygen, nitrogen, carbonic gas, air, etc.
> *temperature*—chemical heat in the blood of warm-blooded animals

electricity—colloidal electricity
magnetism—protein magnetic force and the like
radiation—hydrogen ions, pH phosphorus ions

In studying cells, solids or colloidal gels include the nucleus (nuclear membrane, chromatin, karyosome, and nucleolus), *centriole* (aster), *mitochondria, metaplastic,*[1] and *golgi*. Among the liquids present is cell-sap plasma, and among the gases, vacuole. As cytology teaches us, the internal organization of the cell differs from that of internal organs in the human body in that it is not always visible to the naked eye. During periods of inactivity it vanishes, while during separation or reproduction it often becomes visible.

It is therefore a serious fallacy to think of cells as simple things. The cells of higher animals are complex indeed. We need to be aware of all the hard work entailed in the fact that the entire body of physicochemical laws governing the three states of matter—solid, liquid, and gas—are active within the cell. Crystal bodies are often found within cells, and these crystals can become calcium oxalate or capillary crystal druses.[2]

As we know from cellular physiology, organisms are not produced from a single material but have to meet the difficult conditions necessary to bind them at a single point through the concentration of the solids, liquids, gases, electricity, magnetism, and radiation contained in matter.

The birth of organic compounds

When an organism is born, organic compounds have to be born for the conditions of life to be met. In particular, proteins have to come into being. The reason is that proteins have the capacity to bond solids, liquids, gases, electricity, magnetism, and radiation. Solids, liquids, and gases cannot be joined together in inorganic substances. Melting, freezing, and sublimation points change with the temperature. When organic matter steps on the scene, everything changes.

We must think in terms of a previous arrangement that made the birth of living organisms possible with the appearance of proteins and the bonding of aqueous solutions, starches, fats, sugars, and the like.

It is therefore inconceivable that one combination after the other repeated ad infinitum would result in the birth of a living thing, as H. G.

---

1. See "Cytology," *Collier's Encyclopedia* (New York: Crowell-Collier, 1950).
2. See Léon Hirth and Joseph Stolkowski, *Biologie cellulaire* (Paris: Presses Universitaires de France, 1955).

Wells would have it. This claim is even less feasible when we consider that these living things continue their evolutionary ascent by shutting out the pressures of the environment and selecting their own direction. We have no choice but to suppose a principle of generation from within, much as a flower grows from a seed. I do not mean to deny the pressures of the environment or the place of trial and error, but I find it hard to believe that the fitness of the inner environment of the living organism on its evolutionary path is simply the result of causal or genetic factors added from without.

In the ancient Greece of Democritus, matter was thought of as blind chaos. If we follow modern atomic physics to see the generation of a great, law-regulated, chemically organic world as arranged from the beginning, we may, I believe, consider the birth of living things an a priori probability.

I find nothing unreasonable in Aleksandr Oparin's discovery of *purpose* in the fact that life was born on earth at the time of the stabilization of the solar system, the production of amino acid compounds, the appearance of proteins in 17,000-molecule units, and the approximation of enzyme activity.

But even if life was a *possibility*, living organisms had to be stronger than rock; they had to overcome wind, rain, snow, and lightening; they had to withstand all the threats of erosion from wind, water, and light. It is no easy task to maintain the autonomy of this self-reliance for billions of years. Somehow adaptation of the inner environment proved not to be a complete impossibility for nature's selection principle, which had come up with the plan of making the external environment fit for life.

THE FINALITY OF ENZYMES

The orientation of colloid chemistry

Positive *Sol* absorbs acidic dyes and negative *Sol* absorbs basic dyes. This obliges us to recognize a certain electrochemical design in colloids.

A great many gels and certain kinds of protein take a cylindrical shape. Many atomic bindings stretch out into a cylindrical shape and become isoelectric bodies where + and − exist in balance within a single molecule. One can only marvel at the cosmic design of it all. Moreover, when these proteins are soaked in a liquid with a positive electrical charge such as a basic solvent, they let go of their own negative charge and are

Latent Purpose in the Structure of Organisms

transferred to a positive charge; when they are soaked in a liquid with a negative charge, the protein molecules let go of their positive charge and are transferred into a negative one. The occurrence of this amazing self-adjustment has to be reckoned a miracle of nature.[3]

Colloid chemistry further teaches us that there is a regular order to directionality, as seen in the way the ζ potential works tangentially in the interfacing of liquids while the δ potential works perpendicularly.[4]

H. A. McTaggart's experiments in which he observed that "bubbles in liquid advance in a positive (+) direction" are also of interest here. He attributed this phenomenon largely to the fact that when molecules of oleic acid lie horizontally in a solvent of potassium permanganate, they are on a par with oxidation; and that when vertical, the oxidation slows down. Horizontally, oxidation occurs where the atomic binding within a cylindrically shaped molecule is weak, while its velocity is high, whereas when they are set vertically, oxidation is slower because a single chain of molecules bonded in the form of a two-layered cylinder detaches one band in the chain through oxidation. Here again we see a deep and latent design of great significance in the direction the molecule takes.[5]

Finality in enzymes: lock and key

Nothing proclaims the details of finality in matter as clearly as enzymes. It bears recalling that Oparin began by insisting on a materialistic theory of finality centered on his study of enzymes.

Today enzymes are considered organic compounds produced in the form of crystal substances. Urease was first captured in crystallized form in 1926 by James Sumner, marking the first time an enzyme was caught in a crystalline solid. This fact has a deep relationship to proteins and is necessary for the activity of living organisms. It is as if tens, indeed hundreds of types of matter had been scattered in what seems to be a completely random manner but ultimately were bent to a single purpose as one and only one function is carried out in each part, yielding an altogether amazing and complex display of finality.

As all enzymologists assert, the best way to explain the working of enzymes is through the connection between a lock and a key.

3. See Tamamushi Bun'ichi 玉蟲文一, 『膠質化学』 [Colloid chemistry] (Tokyo: Iwanami Shoten, 1930), 121.
4. Ibid., 72.
5. Ibid., 185.

135

The energy of enzymes

The driving force of enzymes is thought to consist in factors dependent on the transformation of electrons, factors dependent on this transformation,[6] and factors dependent on changes in the atomic valence of the elements of which enzymes are made up.[7] Accordingly, we must suppose that the catalytic activity of enzymes applies to extremely minute electrical activity. Through the changes that take place via electrodes in a single electron or hydrogen ion, an enzyme, without itself changing, achieves its progressive aim of bringing about change in other things.

That this kind of electrical factory in the microscopic realm can carry out its activities with the goal of constructing a factory of applied chemistry which works at a super-velocity to manufacture cells and bacteria is something that exceeds the reach of human intelligence. All of this makes us think that nothing in the universe happens entirely by chance.

The make-up of the enzyme

Enzymes as such are composed of carbon, hydrogen, and nitrogen as well as of the simple organic compound of oxygen, and they contain nonmetal atoms such as phosphorus, sulfur, potassium, and calcium. They also include metals like copper, iron, and arsenic as catalyst atoms.

In few places, it would seem, do these nonmetal and metal atoms display their individuality to the extent they do in enzymes. The electricity of phosphorus always has its place in the nucleus of the cell; iron has an important role to play in the blood of warm-blooded animals, as do copper atoms in the blood of cold-blooded animals. This leads us to think that each of the atoms seems to be outfitted with an original purpose. This is related to what Oparin calls the teleology of matter.

The shape of an enzyme is not necessarily well-organized, but many of them constitute a linkage of ring shapes and chain shapes, easily dissected to disclose their progressive nature.

The progressive nature of enzymes

There is a chemical intermediate within the enzyme known as "diaphorase," from the Greek term meaning "to carry." When animals breathe,

---

6. See Sakaguchi Kin'ichirō 坂口謹一郎, 『酵素』 [Oxygen] (Tokyo: Shūkyōsha Shoin, 1940), 282.

7. Ibid., 241.

diaphorase takes a position in between the form of oxygen and dehydrogenase, so that the enzymes that carry the hydrogen needed for oxygen to be taken in can step in to enable the breathing of oxygen. When that happens, hydrogen is removed through dehydrogenase, which is next in line, leaving only the enzyme to be transported into the body of the animal. Diaphorase is the intermediary transporting hydrogen in this case, but there are numerous things within the enzyme that have this character. Or rather, all enzymes, albeit in different forms, may be thought to have the character of diaphorase.

Sorting in enzymes

The cluster of enzymes that breaks down the peptide bonds of protein molecules and converts them into low-grade peptides with few atoms is known as proteinase. According to the density of the hydrogen ion or pH, proteinase works to effect a general classification of enzymes into three types:

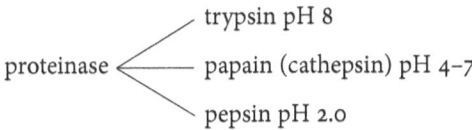

In the systems and cells of animals, pepsin and trypsin are present as the inactive elements pepsinogen and trypsinogen, which are activated through the self-energy of pepsin and trypsin respectively. Trypsin is also present in combination with "inhibitors" or blocking substances. Trypsin and inhibitors are activated through the enterokinase present in pancreatic fluid as the inactive element prokinase, which is then excreted from the pancreas and emerges at the point of intestinal mucosa to become active as enterokinase.

In addition, pancreatic fluid contains a protein catabolic enzyme known as chymotrypsin which is activated by trypsin and has the power to coagulate cow's milk. Within the system it remains in an inactive state as chymotrypsinogen and is activated not by enterokinase and pepsin but by minuscule amounts of trypsin.

It is interesting to note that the location at which proteins are broken down differs according to the type of enzyme. It seems that the enzyme is influenced differently, depending on the type of amino acid that organizes the process. For example, as carboxypeptidase, pepsin breaks down peptides including glutamic acid and tyrosine.

Cosmic Purpose

```
                Carbo-benzoquinone ─┼─ L-tyrosine ─┼─ glycine ─┼─ amino
                                    papain      chymotrypsin  papain
intestinal
  mucosa → prokinase
 enzymes      │
              ▼
       enterokinase → trypsin inhibitors
                              │
                              ▼
                       trypsin → trypsinogen
                                      │
                                      ▼
                               trypsin → chymotrypsinogen
                                               │
                                               ▼
                                          chymotrypsin
```

The arrows (→) indicate enzyme reaction; enzymes are at the left and the basic substances to the right; the broken vertical arrows indicate what has been generated from the reaction and functions as an enzyme.

Carbo-benzoquinone, L-glutamine, pepsin, and L-tyrosine as well as chymotrypsin destroy tyrosine and glycine compounds; papain works like aminopeptidase. In other words, even with the same basic substance, the location of the breakdown differs.

The finality of enzymes

Imagine for a moment that you pull on a cord strung with beads and send them scattering in all directions. Now what if, all on their own, those beads were to string themselves onto a cord. Who of us would not be surprised! And yet this is precisely the kind of astonishing selective activity that occurs in nature as hundreds of enzymes carry out their work among the world's plants and animals. Enzymes are truly incomprehensible without cosmic finality.

GENERATING ELECTRICITY THROUGH PROTEINS

The amino group as an electrical generator

The basic substance known as amino acid has the carboxyl (–COOH) group of acids and the amino (–NH$_2$) or imino (–NH) group of bases. COOH$^-$ has a negative polarity and NH$_3$ a positive polarity. This means they are furnished with at least seven astonishing wonders: (1) a battery, (2) power cables, (3) magnetism, (4) circuitry, (5) a tape recorder, (6) a switchboard, and (7) a power generator with which to make the material that is the basic stuff of a living organism—proteins. Should they not be thought of as energy mechanisms rather than as matter?

Latent Purpose in the Structure of Organisms

This is why, if they can change into hormones by means of these electrical proteins, (1) they can also be the basis for chromosomes; (2) they can become nucleic acid; (3) they can produce nuclear substances; (4) they can be a source of nutrition; (5) they can be dynamic elements; (6) they have the power to generate, and (7) they can serve as energy condensers.

Proteins make possible the following electrical wonders: (1) restructuring, (2) radical selection by way of permeation, (3) selective integration as male or female, (4) generation, (5) growth, (6) regeneration, and (7) life.

Simple materialism looks only at the *form* of the protein to the neglect of the *energy* that runs through it and the direction of that energy. The form is fascinating, but still more interesting is the electric current that runs through the system and the magnetism that shows up at right angles to it.

Proteins are responsible for the restructuring that goes on in the bodies of living organisms. Living things break down proteins into compatible amino acids, join to them, and establish connections to create new proteins. Their ability to accommodate the a priori laws of selection constitutes a particular characteristic of living organisms. This is what Albert Szent-Györgyi came to realize in his study of muscles.[8]

The molecular structure of amino acids

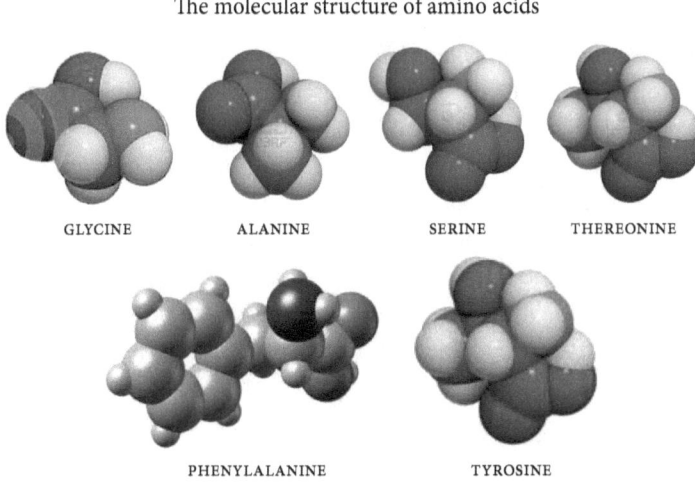

GLYCINE ALANINE SERINE THEREONINE

PHENYLALANINE TYROSINE

8. See his *Nature of Life* (note 15 on page 64 above).

The secret of proteins

It was noticed early on that proteins are indispensable to the preservation of life. In 1838 the agricultural chemist Gerardus Johannes Mulder gave them the name *protein*, which means "the most important factor."

Mulder's understanding of proteins was inadequate, but around 1953, through the efforts of the Cambridge biochemist Frederick Sanger and his team, the chemical structure of insulin, a protein that shapes the pancreatic hormone, was identified. Their research showed that of the 24 amino acids now known to constitute proteins, 17 of them were related to insulin, which in turn is related to 777 atoms, including 254 of carbon, 377 of hydrogen, 65 of nitrogen, 75 of oxygen, and 6 of sulfur.

Sanger went on to show that the unity of amino acids that assemble these 777 atoms form 51 groups which can be lined up in two rows: Row A shows glycine in 21 groups, and row B, phenylalanine in 30 groups. The glycine and phenylalanine rows are connected only in the cystine group (which includes 2 sulfurs). The two rows touch at two places, each with two new sulfur atoms for a total of four; in addition, two cystines in the glycine row are connected to two sulfur atoms.[9]

Of further interest is the fact that amino acids all come in different types, each in the shape of a small battery. Therefore, if the cluster of batteries are lined up in two rows and made into a circuit, a dynamic function is enabled that stimulates hormone growth.

Electrophysiology and finality

In his book *The Phenomena of Life*, which approached the phenomenon of life from a radio-electric standpoint, George Crile published interesting results previously overlooked by physiologists.[10]

A certain student died as a result of a train injury. His wounds were not serious, little blood was lost, and there was no brain damage, which left only the sensational possibility that his death was neurological. The results of Crile's study of this case, as well as studies done on many soldiers taking on the war front in Europe from 1914 to 1918, led him to the conclusion that electrical and radioactive functions have a large role to play in life.

---

9. See E. O. P. Thompson, "The Insulin Molecule," *Scientific American* 192 (1955), 36–41.

10. George Crile, *The Phenomena of Life: A Radio-Electric Interpretation* (London: William Heinemann, 1936).

Latent Purpose in the Structure of Organisms

Crile sacrificed several thousand animals for repeated experiments on this question, noting that there are about ten reasons for the electrical disposition exhibited in life:[11]

1. The electrical nature of water: It is a well-known fact that the water that comprises three-quarters of the human body contains bipolar electricity. This depends on the nature of ionization in water. Further to be noted is that water has a remarkable catalytic function and also contains a particular caloric value.
2. Electrolytic and colloidal chemistry: The greater part of the matter that makes up the human body exhibits a remarkable electrical character.
3. Hydrogen ions penetrate into every corner of the human body and, where concentrated, produce small but fundamental changes. Electrically speaking, these hydrogen ions are truly spectacular.
4. In bundling and solidifying protein-activating nitrogen, proteins created another considerable source of electricity through radioactivity. They show a surprising equipotentiality with the periphery, so that if the periphery is plus, the protein is minus, and vice versa, thus effecting the activities proper to coordination and adjustment.
5. The lipoid membrane that consists of an outer wall of several hundred trillion cells is also very significant for electricity. This thin membrane has the capacity to concentrate high-voltage electrical energy—the thinner the membrane the higher its capacity.
6. The nervous system: In animals, nerves with their ability to serve as conductors of electricity are highly significant.
7. Oxidation begins by attacking the electrons in proteins with ultra-short waves, causing radioactivity. This seems to be how electricity is generated in the human body. These ultra-short waves arise within cellular protoplasm. The electrons thus released bring about nerve activity, concentrate electrical energy in the lipid of the cell walls, and then transmit electric energy to the interior of the cell.
8. Cells are bipolar batteries. Research on amoebas has made it clear that the nucleus of the cell has a positive electrical charge and protoplasm a negative charge.
9. The cell is also a photoelectric tube with the amazing capacity of serving as an electrical generator's power line, an insulator, and a storage battery.
10. Purkinje cells and Nissl granules: Brain cells, particularly in developed Purkinje cells, have a system for increasing the oxidizing activity of

---

11. *Ibid.*, 48.

Nissl granules. Almost like platinum black, sponge-like things capable of absorbing great quantities of oxygen, these Nissl bodies within the Purkinje cells of brain cells have the power, through colloidal chemical activity, to emit the large quantity of oxygen they house, and through that oxidizing activity release nitrogen electrons and bring nerve activity to life. When a cell is destroyed, the Nissl granules are expelled and absorbed into other cells.

The electrical nature of nitrogen in protein cells

As Frank A. Ernst and Mildred S. Sherman of the Fixed Nitrogen Research Laboratory in Washington, D. C. have announced, nitrogen fixation on the surface of the earth amounts to about 100 million tons annually. The 100 million volts of electricity in lightening leaves a trail about a third of a kilometer long and can reach a temperature of 12,000° C. This explosive energy of nitrogen is what makes gunpowder work. Nitrous gunpowder generates ultra-short waves, and Crile observes that the activity of nitrogen in the protoplasmic chemistry of life has the same quality. Protoplasm and nitrous gunpowder:

1. are both carbon compounds;
2. both show rapid oxidation;
3. both generate ultra-short waves;
4. both generate CO_2, free nitrogen, and nitrous compounds;
5. both cause explosions through light waves, electricity, and sound waves; and
6. both show continuous dissociation.

Such are the similarities Crile discovered between protoplasm and nitrous gunpowder.

Catalysis in iron

According to the published results of Otto Warburg:

> Iron is particularly necessary for the bonding of ammonium. In order to bond nitriding iron, active iron and nitrogen are bonded through the catalytic action of iron. This nitride then compounds with hydrogen to make ammonium, and the remaining iron administers the catalysis.[12]

Insects that move their wings more than 300 times per second contain

---

12. *Ibid.*, 71.

large amounts of iron and chrome in the muscles of their wings, offering an extreme example of how the capacity for flight coexists with nitrides. According to Warburg, this enzyme proves to reach localized temperatures of over 5,000° C.

Fluorescence in animals has exactly the same quality, which Crile recognizes as an example of the mystique of enzymes.

Crile further explains that nerve activity in the brain relies on electricity and radiation, differentiating functions within the gray and white matter of the brain. He reports that cells in the brain's white matter that lose their nuclei function like a gramophone recording. Electrons emitted from Purkinje cells call up what been recorded there and then continue their work within the gray matter to display association, judgment, conjecture and other functions.

Crile also strikes on the idea that the activity of the brain has a hormonal connection with the thyroid gland and adrenalin. He notes that the minute elements in the thyroid are imperfectly developed and connected to the development of the thyroid and the brain, so that as the adrenalin gland increases in size, their vitality also increases.

According to Crile, carbohydrate compounds are what account for the influence of ultraviolet rays on plants. In addition, insulin, the hormone in liver, has an important role in oxidizing carbohydrates.

The thyroid in fish, which is no larger than the head of a pin, is attached close to the heart and gills; in lower spinalized animals, it is attached close to the gonads. The thyroid gland supplies nitrogen enrichment to the soil. It is also absolutely necessary for the development of infants, so that a deficiency in thyroid hormones results in deformity. Adrenalin also vitalizes animals in later stages of growth, where it is thought to result mainly from the collaboration of electricity and radiation in the brain.

Pursuing this line, Crile sought to discover the selective activity of physiological functions in the difference of wavelengths in the sun's rays.[13]

Radiation in the brain is cooled through water, so that infants with a large amount of this water coolant show vitality. He observes that adrenalin is great in wild animals, while in humans the thyroid is relatively large.

Atomically speaking, Crile identifies bacteria as nitrous, plants as carbonic, and animals as oxygenic. He further likens nitrous oxygen to a surgical unit that contributes greatly to healing on the front lines where it

13. *Ibid.*, 132.

is administered, without an anaesthetic, to soldiers with external trauma that consume the body's electricity.

Adaptation in physiologically free mechanisms

Life as a free mechanism

With regard to the origins of life, I believe there is a need for a fresh approach to research. Up until now many scientists have made efforts to discover the evolution of life only in terms of concentrations of matter. I do not believe this to be a fruitless direction for research.

However, when I speak of a new direction, we must not exclude the whole collection of laws related to the appearance of life. Oparin has brought the liquid state known as coacervate into the story of the origins of life. A. F. J. Butenandt believed that viruses do not belong to the family of proteins and living organisms, but rather represent a degeneration of the material that makes up cells and hence can only live as parasites.

Life as a concentration of laws

If we look carefully at an amoeba under a microscope, we notice that it is a material phenomenon comprising solids, liquids, gases, ions, and so forth, all concentrated and active within a single cell. Even in the case of a colloid that has become liquid, it is undeniably a solid substance. It falls to the living body to manifest a unified freedom at a fixed temperature. In the world of inorganic chemistry, it is no easy matter to gather together in one place the various kinds of biological laws governing solids, liquids, gases, electricity, ions, magnetism, and radiation. But when it comes to their form as proteins, fats, starches, hormones, and vitamins, each of them are indeed concentrated in one place.

The appearance on the earth of this assemblage of laws seems to completely defy statistic probability. For inorganic chemistry it is totally impossible. Looking only at the differential in the temperatures required for the formation of a solid, a liquid, or a gas, it is impossible to concentrate these variant temperatures on a single point. To make this impossibility possible, things have to be formed into the shape of proteins, fats, starches, enzymes, hormones, and vitamins.

In the world of organic chemistry, solids, liquids, gases, magnetism, electricity, and radiation can all six easily be concentrated at a single point in ordinary temperatures. This points to the presence of an arrangement that makes it possible for an organism to live. The life sciences have so far forgotten these conditions, and indeed H. G. Wells's *The Science of Life* displays this neglect to an extreme.[14] We recall his claim that, given an infinite number of combinations of the letters of the alphabet, the works of Shakespeare could be replicated.[15] If one allowed for an infinite amount of time as well, this might indeed be a possibility. However, even if given an infinite amount of time, would it be possible to take solids, liquids, and gases of different temperatures and concentrate them at a single point? Without some preexisting arrangement, it is altogether impossible to think that something utterly impossible in the realm of inorganic chemistry could be brought about in the realm of organic chemistry.

Similarly, placing proteins, fats, starches, enzymes, vitamins, and hormones in normal temperatures and concentrating them into one small area cannot be considered a chance production. What we have is a concentrated system making ample use of selection. I find it an interesting event in the history of the study of evolution that, on observing this selectivity, Oparin could not but acknowledge a purpose in the world of organic chemistry.

Conditions for mechanisms with a degree of freedom

The intrusion of free mechanisms into the realm of physical chemistry, where everything seems to be determined, that is, introducing the idea of an apparently miraculously variable universe into a process thought to advance by fate, may seem contradictory. Yet nature in no way contradicts the laws of physical chemistry in the intense pleasure it takes in making use of those very same laws to construct the free mechanism of living things. Indeed, it is a marvel to behold the achievement of a structure suitable for freedom of movement, freedom to change directions, freedom to change coordinates, freedom of choice, freedom of growth, freedom to replace laws, and freedom to supervise the division of labor.

Regarding the mechanism of freedom of movement, Szent-Györgyi has shown that the laws regulating liquefaction and solidification become free in virtue of the directionality of bioelectricity. Changes in direction and changes of coordinates become free in virtue of the laws and energy of colloid chemistry.

Through curious combinations at the microscopic level, vitamin pro-

---

14. See note 2 on page 39 above.
15. See page 65 above.

teins exhibit a nearly panacean electricity through a wide range of choices. Making use of surface electricity, osmotic pressure shows selectivity in discriminating such things as selenium, germanium, silicon, cuprous oxide, and the rectifiers of semiconductors found in mercury lamps. Hormones control the freedom of growth and nucleoproteins oversee the division of labor. (Consult an encyclopedia for further details.)

There are four conditions necessary to exhibit adaptations in the degree of freedom of mechanisms:

1. the concentration of the laws of physics and chemistry in a single point;
2. the conditions for structuring that concentration, which include temperature, humidity, atmospheric pressure, electricity, colloidal nature, and so forth;
3. rearrangement at the microscopic level; and
4. the arrangement of patterns of movement.

When these conditions are met, a mechanism is destined to be separated and move in an apparently static fashion or along a single track, thus converting it into a moving body with freedom of action. It is only then that we may speak of the appearance of a "living body."

As I cannot go into further detail on these matters here, let us consider the various mechanisms for the degree of freedom that rely on colloid chemistry, proteins, enzymes, and electrophysiology.

The construction of solid-state mitochondria

Mitochondria, which is present in the cell in a solid state, is a depository of active enzymes essential for higher animals. In the view of Philip Siekevitz,[16] this solid state has an architectural structure that dispenses active enzymes as a source of energy in an order suited to animal development. Nakazawa Tsuneyuki of the Neurological Research Center at Keiō University's Faculty of Medicine has produced a film of his experiments entitled "The Generation of Nerve Fiber" in which the task of mitochondria is clearly featured.

Gamma particles and white blood cells

Besides electrical energy, life needs radiation, as David Gitlin and Charles

---

16. Philip Siekevitz, "Powerhouse of the Cell," *Scientific American* 197 (July 1957): 131–44.

Schematic diagram of a cell

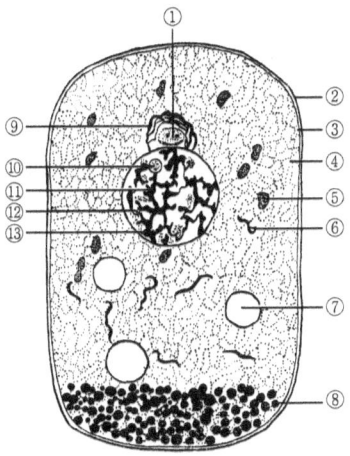

(1) centriole or aster, (2) cell wall, (3) plasma membranes, (4) cortical layer of the cytoplasm, (5) plastids, (6) mitochondria, (7) vacuole, (8) metaplastic bodies, (9) Golgi bodies, (10) nucleolus, (11) basichromatin, (12) linin, (13) karyosome

A. Janeway have described. The Swedish scientist Arne Tiselius divided serum into four types: protein albumin, alpha globulins, beta globulins, and high-level gamma globulins. At certain times infants show a deficiency of gamma globulins. When they contract a disease in this condition, injections of antitoxins cannot precent an inevitable death. Gitlin and Janeway report that as long as gamma globulin protein is transmitted from the mother, the infant is healthy, but when the time comes to produce the protein on its own, some infants lack the capacity to do so. At these times gamma globulins have to be supplemented continuously.[17]

The development of a healthy body is protected by this complicated arrangement, and as a result the provision of the necessary supplements through marriage is assured.

THE PHENOMENON OF LIFE AS SEEN IN MUSCLE ACTIVITY

The activity of ATP

The Hungarian biochemist Szent-Györgyi, to whom we have referred earlier, sought insight into the phenomenon of life by means of a biochemical analysis of muscle activity.

---

17. See David Gitlin and Charles A. Janeway, "Agammaglobulinemia," *Scientific American* 197 (July 1957): 93–105.

Muscle activity begins with the combining of the proteins known as myosin and actin. Their combination produces muscle contraction.

Essentially, life is present in the combination of nucleic acid and protein, by virtue of which things in a liquid state are converted to a solid state.[18] ATP (adenosine triphosphate), which is found in this solid state in myosin, serves as a source of energy for that conversion. Seventy ATP molecules combine with one molecule of myosin, and with each contraction of the muscle, one molecule of ATP is dissociated.

ATP is stored within the skeletal protein in myosin. As actin, it carries a minus electrical charge. By compounding this + myosin and − actin, muscle contraction occurs and one molecule of ATP flies off. When this myosin is examined under an electron microscope, it is seen to have a crystalline structure in the form of minute needles. Liquid fills up the space between the crystals.

During a momentary contraction, the muscle dissolves actin through the contraction of the myosin system and the electrical discharge of ATP and the threadlike fibers are easily broken up into globulins.

But no sooner is one molecule of ATP lost than it is replaced with another molecule of ATP, causing the muscle, with its flexibility for expansion, to relax. Incredibly, ATP has the power to cure spastic stiffening.

This contraction (stiffening) through the electrical discharge of ATP and the transition to a state of relaxation constitute muscle movement. The reason myosin is said to be a physiological cog-wheel is that it influences muscle movement through the operation of ATP protein (the skeletal protein woven into myosin).

The chemical cog-wheel and the law of all-or-nothing

The interesting thing about the study of muscles is that one comes to know how ingenious its mutations are. Szent-Györgyi went so far as to call myosin protein the "chemical cog-wheel" of the living organism.[19] When myosin protein combines with ATP, it effects an "all-or-nothing" adaptability.[20] This in turn lays the ground for the adaptability of ions within the body through procedures more complex than one can imagine.

Research into the tensing and relaxing of muscles has shown the truly startling way in which particular laws are applied within the general

---

18. *Nature of Life*, 36.
19. Ibid., 86.
20. Ibid., 95, 106.

framework of the laws of physics related to solids, fluids, gases, electricity, and radiation and in a versatile response to a particular purpose.

Szent-Györgyi observes in this regard:

> At the creation of life, nature does not make use of new principles, but only reinforces them in accord with particular aims.[21]

The discovery of biological valence

A large number of those in the British school see the struggle for survival as the single greatest cause in evolution and lean towards the view that it is ruled by chance. This tendency runs throughout the thinking of Bertrand Russell.

Russell grounds his view of blind chance in the cosmos on the fact that evolution is a slow process and involves a great deal of futility. I think we have to agree with him on both counts, but the sluggish, often pointless course of evolution does not justify leaping to the conclusion that the cosmos as such is blind and fortuitous. Ludwig Büchner's *Kraft und Stoff*, which was widely read in Darwin's age, is typical of that era in which matter was thought to be a chance production. No scientist today who has learned about the structure of the atom and the remarkable selection principle governing the electron shell will go along with this.

Darwin was further attracted to the ideas of Hermann Helmholtz and his mechanistic view of the life world. Helmholtz's views on the "imperfections of the eye" relied on a unilaterally mechanistic view of nature. He held that evolution progresses with imperfections present throughout the mechanisms of the natural world. I cannot agree that imperfections mean that the mechanisms themselves come about blindly.

In Darwin's time cytology, genetics, sexology, and embryology were not very advanced. In our day the cell has been analyzed biochemically and there are those like Oparin who make claims for finality without forsaking a materialistic standpoint.

The study of chromosomes today is more advanced, as witnessed in the fact that followers of Kihara Hitoshi's thremmatology have been able to create seedless watermelons. The strict standardization and orderliness in chromosomes have become evident. There is no need to deny chance based on changeability, but I hardly think this is all there is to be said, let alone that it represents the essence of the universe.

21. *Ibid.*, 79.

Russell argued against purpose in the cosmos from the slow pace of evolution, but the "stability of species" demonstrates that chance is not the absolute ruler of the world of evolution. The amoeba has not changed for billions of years. The same is true of mollusks and insects. If, as H. G. Wells claims, chance is the fundamental ruler of the life world, it would not have been possible for thousands of species of living things to have kept the same form since the Cambrian period, and it would have been impossible to construct evolutionary theories.

From the time of the Mesozoic formation, throughout tens of millions of years, insects have remained largely unchanged. Shellfish have not evolved for hundreds of millions of years. This, of course, cannot be said of evolution in all species of the life world, only as a general picture.

Life forms exhibit a biochemical arrangement in which not all single-cell organisms evolve and that the privilege of evolution is given to only certain of them. This is no mere blind sifting. As with the atomic valence in nature, I hold that something like a biological valence has been determined in the world of living things. We have to suppose that combinations of this biological valence make their evolution possible.

Things that have exhausted their biological valence are incapable of changing "species" over tens of millions of years. Differentiation takes place, but the fact remains that evolution does not. It is limited to things in which uncounted, unconsumed valences remain.

Biological valence is thus restricted, and we may therefore think that things capable of evolution should be open to it in their chromosomes and that their valence remain undetermined.

The appearance of organizers in living bodies

Finality in the structural elements of living organisms

Hans Spemann of Freiburg University in Germany has laid one of the great cornerstones in cytology with his discovery of organizers in living organisms. Ross G. Harrison of the United States merits mention also for his contributions to embryology. Spemann was awarded the Nobel Prize in 1935 for his discovery of what are known in embryology as "organizers." He carried out repeated experiments on the generation of newts, aquatic animals belonging to the salamander family. Half of the epithelium of the eggs was dark gray, the other half white. At fertilization, the line between white and dark gray shifts and dark gray changes to gray.

Spemann began by tying a newly fertilized newt egg with a strand of human hair in the attempt to split the embryo, only to discover that on the left and the right two complete newts were produced. After this, he tied the split vertically, with the result that while the parts around the stomach—liver, lungs, intestines, and other organs—developed normally, the spinal nervous system as well as regulatory forces did not. He concentrated his attention on this point, concluding that in the first case, an organizer was present whereas in the second, it was absent.

Ectoderm becomes skin, brain, nerves, ears, eyes, and other sense organs. Endoderm changes into lungs, liver, intestines, and other internal organs. Mesoderm develops into other systems such as muscles, blood vessels, the urinary tract, and reproductive organs.

Spemann repeated these experiments with embryos. He then cut away those parts that would become the brain and those that would become the stomach and interchanged them one with the other. He confirmed that in the early stages of development, the brain section that was inserted into the stomach section continued to develop as a brain, and vice versa.

He carried on further experiments with the help of Hilda Pröscholdt, this time using the blastopore of dark gray and white salamander embryos. He took part of the blastopore of the white salamanders and inserted it into the stomach region of the blastopore in the gray ones. He found that the two types of salamander embryo joined to produce a single reptile. He dubbed the cells gathered around the blastopore "organizers," having concluded that their structure concentrated the fundamental elements governing all subsequent development.

The discovery was momentous for embryology and cytology alike. From a teleological standpoint, it marked a major milestone in proving an altogether startling finality.

In later experiments, C. H. Waddington of Cambridge in collaboration with Johannes Holtfreter proved that dead elements in organisms have a role to play in the formation of the nervous system. Further, Joseph Needham of Cambridge and Jean Brachet of Brussels discovered that dyes like methylene blue play a large role in embryology. Not only that, they proved that the stimulus of a platinum needle could provoke parthenogenesis in unfertilized frog eggs. These experiments will probably produce many more interesting results, but even if the chemical structure of embryonic development is to some extent subject to irregular stimuli, one may suppose that it will also be proved that such things do not impede the anticipated development.

The relationship between genes and the organizer of an organism

Oscar E. Schotte of Amherst College interchanged parts of salamander and frog embryos, with the result that the teeth of a salamander showed up in the mouth of the frog, and the jawbone of a frog showed up in the salamander. This makies it clear that the relationship between the genes and the organizers of an organism is variable. It proves that while genes carry elements related to the entire organism into each and every cell, the organism's organizer carries partial structural elements.

There is no denying the important fact that for both genes and organizers a temporal arrangement exists for the development of a finality through a succession of hormones and oxygen. This fact alone should suffice as proof of an a priori finality in the universe.[22]

The selective finality of the protoplasmic membrane

A portion of cell membrane in protoplasm is partially hardened into a high-density protoplasmic membrane. The protoplasmic membrane is always semipermeable and does not allow matter dissolved in water to pass. On the other hand, water passes through completely. That said, the membrane is not always semipermeable. It has been discovered to have an ambiguous finality such that when it comes to the absorption of nutrients into the cell and the elimination of unnecessary material, it can become fully permeable as well as selectively permeable, allowing only certain material to pass through it. The credit for this discovery goes to Wilhelm Pfeffer (1845–1920), a disciple of Julius Sachs (1832–1897), the originator of physiological botany. It was Pfeffer who clarified the phenomenon of "osmosis."[23]

It is worth mentioning here that selectivity in what were thought to be extremely simple protoplasmic membranes is latent. When carefully analyzed, these cell membranes turn out not to be so simple but to entail a complex selectivity with a certain finality. The special attention accorded to this in the history of science also merits mention.

---

22. See Joseph Needham, *Biochemistry and Morphogenesis* (Cambridge: Cambridge University Press, 1942); George W. Gray, "The Organizer," *Scientific American* 197 (November 1957), 79–91; "Embryology," *Encyclopedia Britannica* (New York: C. S. Hammond, 1945).

23. See Numata Makoto 沼田真 and Saitō Kazuo 斎藤一雄,『生物学史——生命の探究』[A history of biology: The study of life] (Tokyo: Chukyō Shuppan, 1952), 87.

ADAPTABILITY IN THE MECHANISM FOR NOURISHMENT

Mother's milk and infant nourishment

A mother's breasts are not "milk bottles." At the moment the infant sucks, it synthesizes the mother's milk. In order to release milk, nourishment is removed from the mother's body and collected for the nourishment of the infant. The mother is not conscious of what is happening, but nature's selectivity works in step with the mother's love. Or perhaps we should say, selectivity provides nourishment with a skill beyond that of the mother.[24]

According to research conducted by Elmer Verner McCollum of Johns Hopkins University, no matter how much vitamin C (ascorbic acid) is given the mother, it does not carry over into her milk. The mother's body adjusts the amount of vitamin C produced in direct proportion to the infant's development. That is to say, studies have found that the amount of ascorbic acid that is added to the mother's milk does not exceed a fixed limit, anything beyond that being expelled into the urine.

The fixed ratio of nutrients

Nutritional chemistry suggests that elements essential for the human body are selectively absorbed in fixed proportions. It has also been shown that what is excreted from the body is also eliminated in fixed proportions.[25]

According to McCollum, on the average the human body excretes 10 grams of salt per day through urine and sweat. In other words, this is the amount of salt that humans can intake.

The results of a study at New York Carnegie Nutrition Laboratory on the metabolism of a man abstaining from food for a month showed that the amount of sodium excreted day by day continues to decreases from 2.5 grams on the first day to 0.1 grams on the tenth day. Even if 10 to 15 grams of sodium chloride is administered, it takes four days for 0.1 grams to be excreted. Clearly the man stored only as much sodium chloride in his body as was needed.

Adults are unable to digest if sufficient amounts of saliva are not

---

24. J. M. Barry, "The Synthesis of Milk," *Scientific American* 197 (October 1957): 121–8.

25. Elmer Verner McCollum, *The Newer Knowledge of Nutrition: The Use of Food for the Preservation of Vitality and Health* (New York: Macmillan, 1928), ch. 8.

mixed with food. Sputum is said to be the best medicine for stomach ailments. When it arrives in the stomach, it aids digestion as various forms of digestive acids appear in succession. This mechanism is not something devised by medical doctors. It was only discovered later by physiologists.

Intestinal selectivity

Carl Ferdinand Cori reports that there is a selection going on in the intake of sugar into intestines that differs according to type of sugar. If the absorption of glucose is set at 100, then galactose = 110, fructose = 43, mannose = 19, xylose = 15, and albino sugar = 9. This demonstrates the intense selection going on in the intestines. Reports indicate that even when sugar types are mixed, the absorption rate for each type does not fall.

The mystery of food instincts

It is surprising to learn from McCollum's research into nutrition that dietary selection in infants determines necessary amounts instinctively. It is worth noting here that the human appetite that resists hunger and thirst is consistent with physiological requirements.

Frogs that live in the deserts of Australia have a system to store water under their skin that allows them easily to survive the long months of the dry season. We can only call this a splendid example of the adaptability of the nutritional system.

Food selection in microorganisms

Microbiologist Nakamura Hiroshi has kindly shown us the selectivity visible in microorganisms and I would like to record his findings here.

The carbon source (c-source) that microorganisms take in as nourishment includes various types of sugars and organic acids, but the particularities of the c-source are of special interest. Consider for instance what happens when stereoisomers are supplied.

Example 1. Glucose and fructose have the same molecular formula, $C_6H_{12}O_6$, but there is a preference for glucose in bacteria and molds; sometimes fructose is not taken in at all. In contrast, there are situations where fructose alone is accepted and glucose completely rejected. This type of selection cannot be explained by the chemical structure of matter. It is as if microorganisms had chosen a goal from the start.

Latent Purpose in the Structure of Organisms

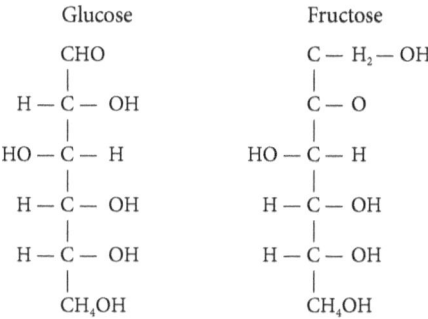

```
        Glucose              Fructose
         CHO                 C — H₂ — OH
          |                    |
     H — C — OH             C — O
          |                    |
    HO — C — H            HO — C — H
          |                    |
     H — C — OH            H — C — OH
          |                    |
     H — C — OH            H — C — OH
          |                    |
        CH₄OH                CH₄OH
```

Example 2. Fumaric acid and maleic acid are isomers with the molecular formula $C_4H_4O_4$, but filamentous bacterium (mold) shows a selectivity towards them. Certain types of mold take in oleic acid as a c-source, but fumaric acid does not use it at all.

```
     Maleic acid              Fumaric acid
   H — C — COOH              H — C — COOH
       ||                        ||
   H — C — COOH           COOH — C — H
       ||                        ||
```

Molecular formula of tartaric acid: CH(OH)COOH
                                     |
                                     CH(OH)COOH

[Tartaric acid stereoisomer structures: HOOC/OH and HOOC/OH (left); HO/COOH and H/COOH (right), both with central OH groups]

In contrast, other kinds of mold take in only fumaric acid and do not use oleic acid. Likewise, this holds even for oleic acid and elaidic acid (an isomer whose molecular formula is $C_{17}H_{33}COOH$).

Example 3. In penicillin (blue mold), glucose and aspergillus (black mold) niger engage in selection when it comes to tartaric acid with optical isomers.

Penicillium glancum aspergillus niger chooses only dextrorotary d–tartaric acid and does not at all accept levorotatory l–tartaric acid. In contrast, bacteria chooses only the levorotatory and rejects dextrorotary tartaric acid.

Example 4. Selection is present where two or more c-sources are given. For example, filamentous bacterium is a c-source that is favorable towards the intake of both glucose and glycerin, but when both are present at the same time, it accepts only glucose and rejects glycerin. This kind of phenomenon completely defies explanation from a chemical standpoint.

Additional examples in higher plants. (1) Seaweed makes a selection between Na and K (closely related chemical elements) contained in the sea. It completely excludes Na, which is present in large amounts, and accepts only K. (2) Even in the same species of flowering plants, we find that rice-plants prefer ammonia while wheat prefers nitrate. Such examples of selection are unexplainable by osmotic theories.

5 Adaptation in the Inner Environment

ADAPTATION IN THE INNER ENVIRONMENT
OF LIVING THINGS

Most biological evolutionists agree that living organisms, which came about as a result of an ingenious adaptability in the natural environment, had the sea as their birthplace. Leaving the sea to step on dry land was an adventure. In order for living things to climb up onto the land, they had to bring with them the adaptability they enjoyed in the sea.

The principle of adaptation in the inner environment of living organisms was discovered by the French physiologist Claude Bernard around 1859, the year Darwin published his *On the Origin of Species*. At first Bernard thought that only blood plasma had this capacity, but in later years he recognized that lymph fluids, along with protoplasm, govern adaptation in the inner environment. In the end, he also included the fluid matrix of the circulatory system in the picture. Thus he came to insist on the homeostasis of water, oxygen, uniform temperature and nutrition (salts, fats, and sugars), and the like.[1]

The terminal use of entropy

As I have discussed in an earlier chapter on the fitness of the environment, the world of matter may be thought to exist so that life might emerge.

Concerning energy, one of whose aims is the appearance of life, the atomic physicist Erwin Schrödinger held that in order to preserve the energy in wave motion (which is eventually extinguished), it is necessary to convert atoms with wave motion into matter.

---

1. Claude Bernard, *Leçons sur les phénomènes de la vie communs aux animaux et aux végétaux* (Paris: Baillière, 1878–1879).

For the energy preserved in atomic form to appear in life form, the laws that pervade the world of matter needed to be concentrated at one point, and that point needs to be given permanence. In the inorganic realm, differences in temperature separate out solids, liquids, and gases without concentrating them in a single point. This only happens when we come to the organic realm, where solids, liquids, gases, electricity, magnetism, and radiation come together. It is when the inorganic passes over into the organic that the creative evolution of the earth becomes really interesting.

Many biologists hold that the structure of an organic body and the alliance of solids, liquids, and gases first became possible in the waters of the earth's oceans. When we add to this the necessary interventions of electricity, magnetism, and radiation, there is simply no way to think, as H. G. Wells does in *The Science of Life*, that all of this was a "chance" construction or that the laws behind this construct and their perpetual continuity are the result of chance. If that were so, chance would be the creator of the laws.

Since the discovery of solids, liquids, gases, electricity, magnetism, radiation, and so forth gathering in a single point through the formation of organic bodies, it is not unreasonable to think of the adventurous voyage from the sea to dry land and the skies as a plan.

Claude Bernard's theory

But there is one necessary condition here. The adaptability of organisms to sea water had to be internalized in the living organism, surmounting external changes, overcoming internal difficulties, correcting, and replenishing. Claude Bernard's attention to this fact uncovered a new dimension to physiology that had escaped Darwin.

Faced with their enemies on dry land, living organisms found it necessary to attune their blood and internal fluids. This required achieving levels of near uniformity in such things as (1) water, (2) fats, (3) proteins, (4) salts, (5) calcium, (6) heat, (7) oxygen, (8) electricity, and (9) hydrogen ions. This surprising homeostasis and tuning constitute a beautiful and indispensable law for the appearance of life.

The problem of uniformity in the density of hydrogen ions

The foods set daily on our tables include salty and sour things. By consuming them, the density of hydrogen ions in the human body, or pH, is kept at a steady 7.4. The number 7.4 refers to the density of an ion measured with a denominator of 1 divided by 10,000,000.

In this minutest of realms a buffering action is at work to keep the body in a weak alkaline state. This can only mean that the adaptability of the inner environment of an organism is more ingenious than we can imagine. As Koga Shōzō notes, "The bodily fluids in humans and other higher animals generally form a single buffering fluid to make adjustments preventing serious changes in pH."[2]

This buffering action depends mainly on a balance of carbonates and bicarbonates in blood plasma. When we think of the way in which harmful carbon gas in the human body regulates the pH in blood to a level of 7.4, we are amazed at the adaptational structure of the body's inner environment.

The strict density of hydrogen ions

Warm-blooded animals stay connected to life at temperatures between 35° and a high of 41°. But the density pH (hydrogen ions) that neutralizes plus and minus ions and allows for the continuance of life admits only of what, electrically speaking, is a truly minimal margin of between 7.3 and 7.5.

Hence, beginning with the density of hydrogen ions, we understand how strict the adaptational conditions for survival are, despite the wide range allowed for extreme variations in the struggle for survival, in atmospheric changes, and in modifications in the food supply.

It takes great effort to effect a buffering action to maintain the density of hydrogen ions between 7.3 and 7.5 so as to regulate the addition of phosphorous into cells, the work of bicarbonates, the use of proteins, or the free use of chromoproteins containing iron. The successful maintenance of pH density is something of a miracle.

Taking into account this achievement and the strict regulatory adaptation it entails, I find it hard to subscribe to the idea of chance sifting advanced by Darwin, Huxley, and others. Quite to the contrary, I firmly concur with Lawrence Henderson's idea of the "fitness of the environment."

I am one of those who believe that without a final purposiveness aimed at life, these adjustments would be absolutely impossible to explain.

---

2. Koga Shōzō 古賀正三, 『pH概説』 [An outline of pH] (Tokyo: Kyōritsu Shuppan, 1953), 172.

Cosmic Purpose

Blood

We must not overlook the fact that the work of blood is connected to colloids and crystals. In startling fashion, hemoglobin in the form of iron colloids supplies oxygen, and by being breathed into the lungs, carries this oxygen into blood vessels where it is distributed throughout the body.

The same can be said of blood molecules. And let us not forget that water is supplied to the human body through colloids in the blood. Water within the cells of higher animals is introduced through the blood.

As for its ingredients, the blood may be said to include all the chemical elements and molecules that make up a living organism. Among the important ones are the eleven types of light atoms with their roughly forty-five microscopic activating elements. It is a marvel how blood generates, repairs, preserves, and regenerates all the organs of higher animals. I yield to medical experts like Amano Shigeyasu when it comes to details of blood research,[3] but would only draw attention to the surprising phenomenon of what Walter Bradford Cannon has called *homeostasis*.

Blood not only generates, repairs, preserves, and regenerates the living body, it carries out functions not ordinarily seen in fluids, such as maintaining body temperature, regulating density levels of things like hydrogen ions, normalizing the functioning of permeability, and resisting pathogens.

Some claim that blood accounts for one-tenth of body weight[4] and others, one-thirteenth.[5]

Finality in red blood cells

Amano reports affirmatively that there is a finality in red blood cells:

> The endowment of vertebrates with red blood cells can be explained as a finality aimed at a complex organization to prevent a random supply of oxygen. A special, closed blood-vascular system is perfected as a pathway for red blood cells. Insects, phoronids, and the like buried in muddy seas and mire also have red blood cells, and even in their case we find specialized closed blood vessels for red blood cells.[6]

---

3. Amano Shigeyasu 天野重安,『血液学の基礎』[Fundamentals of hematology] (Tokyo: Maruzen Shuppan, 1948).

4. See *Collier's Encyclopedia* (New York: Crowell-Collier, 1950).

5. See 『大日本百科事典』[Japanese encyclopedia] (Tokyo: Heibonsha, 1934).

6. *Fundamentals of Hematology*, 145.

Adaptation in the Inner Environment

Finality appearing in antimetabolites

Dilworth Wayne Woolley of the Rockefeller Institute for Medical Research writes in his celebrated work on the study of antimetabolites, that "an understanding of why selectivity exists is important... to satisfy curiosity about natural phenomena."[7] The age has passed for explaining all physiological functions mechanically. From the 1920s, selection theory has permeated the field of biochemistry.

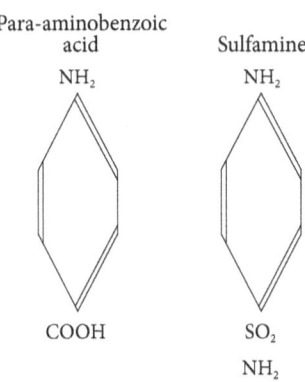

As the study of anti-vitamins antagonistic to vitamins, anti-hormones antagonistic to hormones, and so forth advanced, numerous counter-examples have emerged, but it is clear that a selectivity exists that would not occur without similar chemical structures.

Para-aminobenzoic acid and sulfamine are structurally comparable. As a result, experiments have shown that no matter how much of the sulfuric drug sulfamine is added, it is of no use on its own against bacteria like those com-pounded of para-aminobenzoic acid. Woolley has reported in detail on how these structurally similar antimetabolites oppose metabolic elements necessary to life. There are some sixty examples of this. There are also reports of recovery from the repression of reproduction in microorganisms by pyrithiamine through the use of the similarly structured thiamine and of methylenebis antagonistic to vitamin K.

Of course, such selection as well as selection with regard to antimetabolites varies according to the type of organism. Differences in selection appear in the competition between metabolites and oxygen, and varies according to the strength of the resisting matter to destroy metabolites. Moreover, the functioning of the organism differs according to the physical selection in antagonistic metabolites, and variations in strength appear according to the capacity to adhere to the their physiological structure.

While ketone (CO) is antagonistic to the human metabolism, its affinity to human hemoglobin is two hundred times stronger than its affinity to O_2, but only 0.7 times stronger in the invertebrate parasite gastrophilus

---

7. Dilworth Wayne Woolley, *A Study of Antimetabolites* (New York: John Wiley & Sons, 1952), 147.

intestinalis. It is reported that in addition to differences in the type of organism, proteins related to the transporting of O_2 do not have the same strength of affinity towards antimetabolites but are open to selection.[8]

Dürken's teleology of life

Dürken's holistic teleology

After Charles Darwin, the tendency to deny finality in the life world grew stronger, yielding to casualism and mechanistic views of life.

However, Darwin himself did not necessarily abandon a teleological view of life. Despite the strong accent that *On the Origin of Species* places on life as blind, in *The Descent of Man* Darwin was just as insistent on the choice of mates, using language that even affirmed a notion of a "creator" bordering on teleology.

In a late letter to a Swiss student Darwin states that he had not completely thrown out the idea of "finality." His friend Alfred Russel Wallace argued for an evolutionary theory with a finalistic view of the cosmos. Later studies in genetics and the like by Gregor Mendel, Thomas H. Morgan, and others led to a correction of the view of life as blind, with theories of mutation being propounded by Hugo de Vries, William Bateson, and Andrzej Kozinski posing strong opposition to Darwin's tendency to neglect finality.

A major milestone was reached with Hans Driesch's *Philosophie des Organischen*. His research focused less on the dynamics of generation than on patterns in the generation of life, leading him to advocate a triple harmony in the surprising efficiency of organisms in the life world: (1) the harmony between the whole and the parts, (2) the harmony among the parts, and (3) the finality of the whole.

Driesch located this harmony in the entelechy (purposive competency of function) present within organisms. The French philosopher Henri Bergson and the Irish playwright George Bernard Shaw were among the influential figures who agreed with this purposive vitalism.

German, British, and American biologists, meantime, opposed Driesch's vitalism as too philosophical. This led to the birth of a holistic biology which was neither a mechanistic nor a vitalistic views of life, but set out to judge everything critically from a holistic standpoint.

8. *Ibid.*, 159.

The beginning of holistic biology

If memory serves me, the first in the British Commonwealth to proclaim a holistic biology was Jan Smuts, a general commanding the British army in South Africa. Playing on the English word *whole*, he coined the term *holism*. He stressed that living bodies should be studied as a whole not merely by looking to their individual mechanical parts.

Smuts's holism lacked the kind of strong empirical basis insisted on by the German biologist Bernhard Dürken in his study of developmental dynamics. For details of his experiments, I defer to Dürken's book[9] in order to focus here on trying to understand his turn from a mechanistic theory of life and the finalistic vitalism of his predecessor Driesch to a holistic view of life.

For Dürken, the mechanistic view of life is (1) analytic, (2) deterministic, and therefore (3) mechanistic; it is further (4) blind and (5) ontogenetically mistaken in its insistence first on differentiation and secondly on integration. Similarly, he rejects vitalistic theory on the grounds that it (1) is integral, (2) recognizes a transdifferentiation in opposition to determinism, (3) stresses external energy (entelechy) as an overall guiding factor, (4) denies quantity, (5) like mechanism, stresses first differentiation and secondly integration, (6) goes too far beyond biology into philosophy, (7) is therefore incapable of explaining change as a whole, and (8) cannot explain fixed drives by means of entelechy.

The truth of his own holistic view of life lies in the fact that (1) it puts integration first and differentiation second, (2) it does not shift in the least from Driesch's insistence on affirming the existence of internal elements guiding the whole, (3) in contrast to vitalism remains within the confines of biology and guards against becoming philosophical, and (4) unlike Driesch, gives very serious consideration to quantity.[10]

While avoiding teleology, in the end Dürken cannot avoid accepting finality to explain developmental stages that represent parts segmented from the whole:

> The concept of a primary wholeness clearly includes concepts of internal harmony, orderly arrangement, and the function of self-preservation. In the instinctual movements of an already complete organism, elements of general direction, self-preservation, and finality are clear,

---

9. Bernhard Dürken, *Entwicklungsbiologie und Ganzheit. Ein Beitrag zur Neugestaltung des Weltbildes* (Leipzig: B. G. Teubner, 1936).

10. *Ibid.*, 160.

but these things are especially clear in its development. Suffice it to recall what happens when the initial materials in an undetermined germ diminish, change, and increase. The adjustments that appear at these times exhibit nothing less than the clear presence of an organizing function of the whole which cannot be destroyed in the process. As the parts are finally determined and as long as they do not damage the aspect of the germ that has been determined—indeed, even in the case of a single part at a later stage—orderly arrangement, self-preservation, and therefore finality of action are present together with its primary wholeness. Accordingly, even in an individual in its complete form, the parts and organs as well as the apparatus of the germ are in mutual harmony. With perfect harmony and mutual adaptation they make an integrated system, in such a way that each of the elements that make it up, whatever their role or however differentiated they are, can work together for the whole to continue in existence, leaving us more than a little surprised.[11]

Dürken attacked Driesch's theory of finalistic energy on several counts, but it bears mentioning that when it came to the question of the orderly arrangement of generation, he adopted the exact same arguments.[12] While rejecting finalistic vitalism, he took up a finalistic holism. Having already accepted purpose in biology, he cannot be said to be very far from the theories of Driesch that viewed the extension of finality in differentiation as entelechy.

In putting the accent exclusively on the dynamics of generation in his treatment of wholeness, Dürken found himself struggling when it came to justifying the role of chromosomes.

In my view, the wholeness that Dürken emphasized actually carries all the chromosomes in the cells of the body, so that wholeness reappears even if part of the one-fourth present at the early stage of cell generation or the one-sixteenth present at the stage of cell division is removed. I do not find this all that surprising. Through the chromosomes, each one of the divided cells is a carrier of the whole. A careful reading of Dürken's book leads me to the curious conclusion that, due to an excessive emphasis on the dynamic side of generation, he failed to see the remarkable contribution that the chromosomes make to the whole during the early stage of generation.

In fact, nothing in all of life is so surprising as the chromosomes. The

11. *Ibid.*, 129.
12. *Ibid.*, 160-1.

Adaptation in the Inner Environment

hundreds of millions if not trillions of cells that make up the individual each store the key to the whole. This may be called entelechy.

I feel it necessary to reaffirm the fact that in addition to its other contributions, holistic biology lends strong support to the existence of finality in the organic world.

From his holistic standpoint, Dürken shows a certain degree of timidity when it comes to chromosomal science,[13] which is not opposed to holistic biology but rather, in my view, adds new knowledge to the structure of the biological individual as a whole with its recognition of finality in surprising places.

Holistic biology is diametrically opposed to chromosomal science in its orientation, but there seems to me a danger here of affirming wholeness only in quantitative terms. Dürken admits to a range of predetermination in generation as well as a qualitative integration of whole and parts, but I think it would be better to put a bit more stress on the finality of the whole, including spatial continuity and temporal progression. Chromosome research is not his field, but there is no way to speak of *wholeness* without it. I think further attention is due the amazing finality in the adding, subtracting, multiplying, and dividing of chromosomes, and in the changes taking place within eggs and sperm.

Without too much difficulty, a solution is at hand to holistic biology's problem of quantity and chromosomes, to the problem of protoplasm that Dürken adhered to, and to the problem of the primary and secondary by making a fresh start from Aristotle's teleology. (I admit the scientific basis of the theories of the Russian biologist Trofim Denisovich Lysenko on the role of protoplasm in genetic material, but he went too far in rejecting chromosomes out of hand.)

In order to realize the purpose of an individual, an assembly in the form of segmentation is also needed, as is a progressive development. Without an appreciation of teleology how can such assemblage and progression be explained? Mechanical views of life deny immanent purpose, neovitalism affirms immanent purpose in each part of the cell, and holistic generational dynamics makes their integration primary. As for those who, in arguing for a holistic biology, deny finality outright, I would reckon them as having fallen back into a neomechanistic position. I find the complete failure of H. G. Wells's *The Science of Life* on this count deplorable. After hundreds or even thousands of descriptions

13. Ibid., 131–51.

of the amazing adaptability of the biological world, he has the nerve to conclude that it is the result of chance.

Can there be *chance* with *adaptability*? Is there chance that includes sifting and selection? Is chance with law and regulation possible? The logic of Wells is so riddled with contradiction that one can only stand dumbfounded and shocked at its audacity. Those who engage in science need to be a bit more honest. Standing before the marvelous existence of nature, must we not declare our willingness to renounce prejudice and learn humbly?

I find it interesting that in his attempt to grasp the whole—a point on which he differs from Wells—Dürken also examined the struggle for survival from the same viewpoint to claim the presence of a biological balance.

Adaptation in the genetic mechanisms

Selectivity and adaptability in genetic material

The chromosomes in sex cells carry out addition, subtraction, multiplication, and division. The same holds for cell mitosis. Nothing is quite so startling. Could there be selection at work behind it?

In October 1954, F. H. C. Crick of Cambridge University published a paper in the *Scientific American* in which he explained the secret to us. I will draw on it to examine biological selection in chromosomes.

Chromosomes are made up of (1) protein, (2) deoxyribonucleic acid or DNA, and (3) ribonucleic acid or RNA. Crick does not touch on RNA, focusing rather on DNA in order to explain an astonishing selectivity.

Crick holds that there are two perfectly fitting base pairs: the interaction between adenine and thymine, and the interaction between guanine and cytosine. (See the diagram of the molecular formula of DNA on the facing page.)

He enumerates the conditions for this interaction: (1) it is a chain of phosphorous and sugar having a basic nature, (2) the pair is arranged on a horizontal plane, (3) the order of the arrangement is precisely maintained, (4) the distance between the molecules is equal, and (5) it is subject to rules.

Not only must the selection go smoothly, but the fit must be precise and without any error in measurement—hardly a task for chance. (1) Although alignment on a horizontal plane is difficult, additional care

Adaptation in the Inner Environment

must be taken that the selection of (2) distance, (3) order, (4) rules, and (5) homogeneity are also perfectly arranged. This is detailed work, like sorting currency according to value. It must conform to a precise target, set at a point measuring about one one-hundred-millionth of a centimeter. Hence natural selection is in no sense "haphazard"; it can only be thought of as the adaptation of a finality whose goal is set from the start.

Adolf Butenandt also points to the amazing capacity for adaptation in genetic mechanisms. In a lecture delivered on April 1955 at Tokyo University, he hypothesized that the secret of heredity lies hidden in nucleic acids (Deoxyribonucleic acid or DNA) in the genetic matter within the nucleus of the cell.[14] Research on the reproduction of bacteriophage has made it

---

14. [Kagawa cites a work in Japanese that could not be located. The text of the lecture referred to is, however, available in the original German and Japanese translation: *Vorträge von Professor Dr. A. Butenandt in Japan*, vorvereitet von der Japanischen

167

Cosmic Purpose

clear that nucleic acid is more important than proteins. In the normal state of a cell, it has become clear that normal metabolism depends on the overall integration of nucleic acids. Crick and his co-worker James D. Watson have determined that the nature of the reproduction of a microstructure keeps DNA hidden within the filamentous structure.

DNA is intertwined like a spiral coil. Two strands of thread are bound together as an intermediary bridge. One side of the bridge is long and the other short, making a tight fit. It is said that that the thread is made of pyrimidine and purine bases.

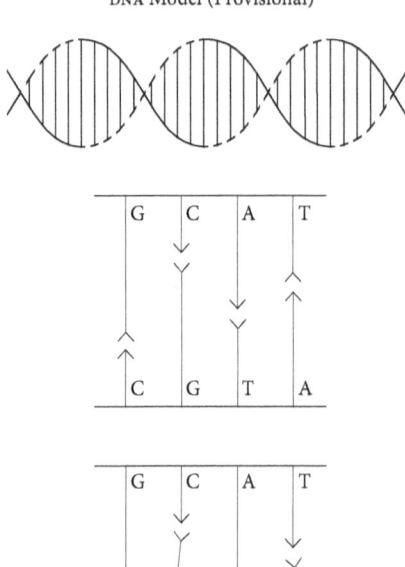

DNA Model (Provisional)

These compatible threads are thought to have a plus-minus relationship.

If the two strands of thread are separated within the cell, one of them is thought to be a prototype for synthesizing the DNA threads, similar to the way letters are imprinted on a coin.

According to Butenandt, the genetic factors supply the cell with characteristic proteins and nucleic acid becomes the controlling core of the process.

In an essay on the biochemistry of the gene function as an example of tryptophan metabolism, Butenandt draws attention further to the collaboration of enzymes forming the greater part of protoplasm. He notes that genes deal with particular enzymes.

Genes hold a fixed position within the cell's nucleus. Genes are necessary to produce a certain ommochrome (visual pigment) in the drosophila (fruit fly). But should a sudden mutation bring about a lack of even a single fixed precursor of tryptophan metabolite, ommochrome is not synthesized. These precursors include the generation of kynurenine from

---

Biochemischen Gesellschaft bei der Einladung zum XIV. Japanischen Medizinischen Kongress in Kyoto (1955).]

Adenine Thymine Cytosine Guanine

tryptophan and then oxykynurenine, as well as the synthesizing of ommochrome.

This chain reaction is catalyzed by characteristic enzymes that function in each reaction process and carry it along.

When these genetic factors adapt, electrons are polymerized one by one with other electrons, atoms with other atoms, and molecules with other molecules, adapting to form several thin, mica-like slices. We should note here the organizational entity behind this selective finality.

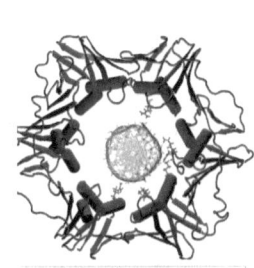

The structure of DNA

As Crick reported in his 1954 article in *Scientific American*, the principal component in the molecular formula of genes is a base substrate. Sugars and phosphorous compounds forming the molecular group of adenine, thymine, cytosine, and guanine are connected and reconnected about a thousand times.

The selection involved in joining only these four as a molecular group borders on the miraculous. The distance between these molecules is said to be a mere 1.5 angstroms (one angstrom, as we recall, being one hundred-millionth of a centimeter). This forms a spiral wrapped about a uniform axis. Crick explained this with x-ray photographs, stretching the circular DNA out like a string.

He also presented a hypothesis on male-female polymerization.

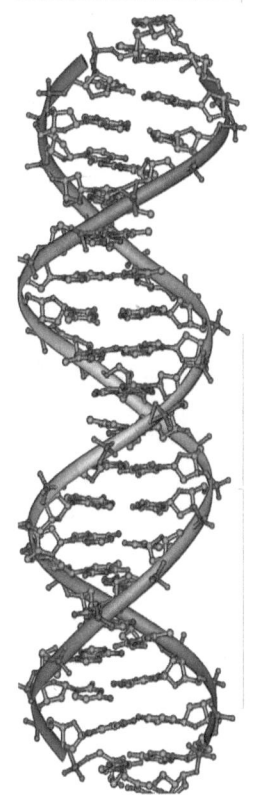

Although the genetic string that structures chromosomes is spiral, it is thought that when cytosine is attached from without to the place held by adenine, the two are coupled, as is guanine with thymine and thymine with guanine. This makes it easy to produce things like grandchildren.

The chemical basis of genetic material or stairs of chemical evolution

Melvin Calvin divided the chemical evolution of atoms on earth into five processes:

1. When an aqueous solution containing formic acid, formaldehyde (a simple carbon compound), and nitrogens like ammonium, nitric acid, and nitrogen are exposed to the influence of the ultraviolet rays in sunlight, amino acids and heterocyclic compounds are produced.[15]
2. As the earth cooled, metal carbides appeared, producing acetylene, which, catalyzed by rocks and minerals, was polymerized into chains.[16]
3. Matter containing radioactive material and two carbon atoms from cosmic rays were born as molecules containing four carbon atoms like succinic acid. These molecules are the metabolic products of important organs in living organisms.

$$\text{Succinic acid}$$
$$\underset{\underset{H\ \ H}{|\ \ |}}{HO - C - \overset{\overset{O\ \ \ H}{|\ \ \ |}}{C} - \overset{\overset{H-O}{|\ \ \ |}}{C} - C - OH}$$

4. If an electrical discharge like lightning is used on methane, hydrogen, ammonium, water, and the like, chemical compounds of carbon atoms are generated along with the amino acid compound forms needed for protein.

With the appearance of catalytic material, chemical affinity evolved at a speed of ten million times in one ten-thousandth of a second, and as glycine and then phenylglycine were produced from methylamine, molecules grew more complex. The larger molecules got larger, the more catalytic action increased.

Aromatic compounds, which are collections of flat molecules, crystallize in the form of a stack of cards. According to their position in the alignment, the atoms and electrons within them are pulled together onto

---

15. Suggested by publications of the Japanese Biochemical Society.
16. Suggested by Oparin.

Adaptation in the Inner Environment

the same plane. This is the structure of DNA that was confirmed a number of yeas ago. As Calvin explains, this represents the genetic profile of life.

Adaptation in the genetic material "nucleic acid"

Crick has studied in detail the finality of nucleic acid. I am mainly following his work here.

There are two kinds of nucleic acid: deoxyribonucleic acid, normally referred to as DNA, and ribonucleic acid, or RNA. DNA is present only in the nucleus of the cell, while RNA is found in protoplasm. Things having DNA and RNA are combinations of phosphorous and sugar, while the parts that differ somewhat belong to four types of bases. Deoxyribose, adenine, guanine, and cytosine also adhere in RNA, but with DNA a fourth, thymine, is added. In the case of RNA, uracil is added. There is no change at the point where the central elements phosphorous and sugar bond, but the lateral position at which the four elements of adenine, guanine, cytosine, and thymine and so forth are attached is disorderly. Much the same as letters are arranged differently according to the word being spelled, scholars attach profound meaning to the point at which no fixed order is maintained.

Two strands of spiral matter

What appears under an electron microscope to be a single strand of thread turns out to be two. Their combination is in no sense a matter of chance. Guanine is tied to cytosine and adenine is connected to thymine.

RNA is not as easy to understand as DNA. Nevertheless, adenine alone is wrapped up with adenine and the uracil group only wound with the uracil group. It is also thought that when adenine is tied up with uracil, hydrogen helps them unite. We may further note that Alexander Rich reports having discovered a spiral coil of three strands.

Arthur Kornberg of Washington University in St. Louis succeeded in an interesting experiment aimed at extracting the four groups of molecules attached to DNA from within cells as living DNA. He reported two conditions: (1) each of the four groups must be present and (2) DNA is present as a seed.

A hand-in-glove fit

Studies by Cyrus Levinthal of Michigan University produced interesting results regarding a virus found within bacteria and known as T$_2$. He established that the DNA of the virus was not a single unit but was made up of several groups. Within the virus there are gigantic elements with 12 million molecules, compared to which the other elements are small. Levinthal investigated them using radioactive phosphorus atoms and came up with a surprising result. Half of the DNA in a newly replicated virus had no radiation. Still more interesting was the fact that when the DNA replicated, it resulted in radiation-free DNA. From this we learn that while half of that DNA is produced from a parent, the other half is completely new. For Crick, this means that the genetic material in DNA fits like a hand to a glove. In other words, we might say that both glove and hand are included in the gigantic DNA within the genes of the virus, and that when hand and glove are detached in replication, the hand again fits into the glove, while the glove sets out in search of a new hand.

Crick recognized that this formulation showed certain differences from what Gunther S. Stent of the University of California, Berkeley had found in his research. On the other hand, Columbia University's J. Herbert Taylor and his collaborators backed up Crick's "hand-in-glove" theory when they found, while searching for DNA in elements with radioactive heavy hydrogen (deuterium), that half of what is replicated is radioactive and the other half is not.

The power of RNA

Heinz Fraenkel-Conrat of Columbia University and Gerhard Schramm of Tübingen University recently made an interesting discovery in studying the tobacco mosaic virus. They isolated RNA from the protein of the virus, which they then mixed with other proteins and injected it into tobacco leaves. What they found is that the RNA picked out what suited it from the new protein material to create the same form of virus. This shows that RNA creates new proteins. Crick had claimed that nucleic acid takes the initiative in genetic material. Even if his theory has not won universal approval in the scientific world, we may recognize it as extremely influential. It is striking to see the study of genetic factors come this far.

Adaptation in the male-female mechanism

Nothing is as complex as research into the "sexuality" of algae. Why it should be so complicated is completely beyond me, but I do understand that the mechanism of sexuality is no simple matter.

As we see in the simplest green algae (chlorella), algae cells apparently cluster all on their own. Examining the sexuality entailed, one is struck by the mystery of why this should occur in such a low life form. Reproduction in algae takes place through the division of nutritional cells into two or through the division of their contents, through aplanospores and autospores, or else through autocolony.

Ulvaceae, a different variety of green algae, is made of one or two layers of cells and has a thin flat or tubular form. A single strand of this shape may diverge or divide. Asexual reproduction depends on akinetes and autospores having two or four strands of cilia. Akinetes are detached from the edge of the main body. But when part of this "body" is cut off, it continues to grow.

Sexual reproduction takes place in alternating generations. That is, one generation reproduces asexually, the next sexually, and the third asexually again.

Sexual reproduction depends on the union of gametes having two strands of cilia. Depending on the type, the female in these gametes (or zygotes) is often larger than the male.[17]

In brown algae, when the trophozoite is cut, only a few reproduce immediately, but even so, the fact is that some of them do. This type makes twigs or "hatching buds" (*Brutknospen*) that become new individuals or sphacelariae. As a characteristic of brown algae (phaeophyceae or kjellm), eggs as well as reproductive cells are generally all mobile. Reproductive cells by and large have one reddish-brown eyespot, to whose horizontal axis the cilia are attached.

Reproduction is asexual and sexual, the former depending on zoospores which are spores with the power of movement. Spores are formed when the contents of the mother cell break up. These mother cells are referred to as zoosporangia or unilocular sporangia.

The asexual spores involved in asexual reproduction are not mobile and come in two types: those in which an individual cell divides into four parts and those in which individuals cells have four nuclei. They are immobile. Thus we have reproduction through the union of two mobile individuals into a gamete and reproduction through sperm and the fertilization of an egg.

Gametes are divided between those of the same size, called isogametes, and those of different sizes, called anisogametes. An egg, or female, is

---

17. See Okamura Kintarō 岡村金太郎, ed.,『日本海藻誌』[Japanese algae] (Tokyo: Uchida Rōkakuho), 1936.

much larger than the male sperm. It is also colored. This mother cell breaks up into several cells, each of which generate a spore called a gametangium or plurilocular sporangium. Many of these create clusters of reproductive cells or sori.

Reproductive cells are made from cortex. Some of them, like fucaceae of the fucales family, a species of edible brown seaweed, generate reproductive cells in a special subcortical hollow known as a conceptacle. Since alternating generation is frequent among brown algae, this is included in their classification. Within the conceptacle a gender-neutral filament (paraphyses) is produced, at whose roots the male and female reproductive elements appear.

The bisexual mechanism of algae is wonder enough, but when we examine the evolution of cryptogamic plants and flowering plants, the complexity of the mechanism is overwhelming. As John Merle Coulter says, we can only suppose that the aim is not just to continue but to evolve.

Algae migrated to dry land as lichen, became ferns, evolved into gymnospermous plants like ginkgos, and in certain plants took the form of distinct male and female bodies. When this was found to be inconvenient, flowering plants took the form of gathering several hundreds, even several thousands of flowers into one, as in petaled flowers like the chrysanthemum. By means of such petalage, plants suddenly displayed a potent means of survival so that life forms suited to survive on the coastline could advance to desert regions not inhabited by other life forms.

Not to lose out to the discovery by plants of a bisexual mechanism for adaptation, animals also worked out a way to evolve bisexually. (See illustrations of the evolution of flowers and plants on the following pages.)

Arthropods invading dry land as insects made what remains, past and present, a rare and momentous discovery. Devising ways to move from the water directly to flying in the sky, and through metamorphosis to seclude themselves in the earth where they found a means to survive the cold, insects perfected great inventions that surpass the understanding of human intelligence. The thought of the insect nervous system alone always leaves me dumbfounded.

Ants and bees divide their labors socially. This also entails physiological determinations. The number of males is limited and the queen is the reproductive specialist. In these insect societies a social language is employed and controls are enforced on insect behavior. Throughout tens of millions of years customs were preserved that make insect societies a marvel of survival even within the animal world. It is said that those

Adaptation in the Inner Environment

The Evolution of Flowers

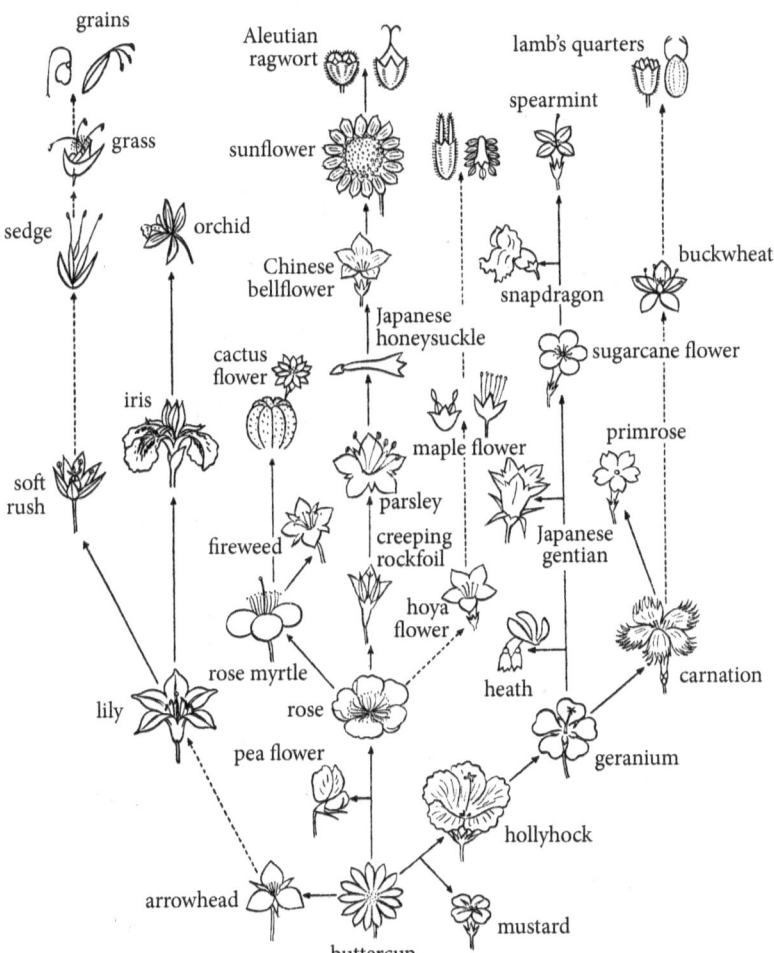

cultivating coffee plantations in Brazil enter into a competition with the ants to see which side will win. To study the social activity of agricultural ants is to be astonished at their organizational efforts.

The finality of the bisexual mechanism

The mechanism of plant sexuality differs somewhat from that of animals. In recent years research into sexuality has made advances around the world. The leading botanist at the University of Chicago, J. M. Coulter, to whom we referred above, has published a work on *The Evolution of Sex in Plants*. In his research on the sexuality of spirogyra published in

Cosmic Purpose

The Evolution of Plants

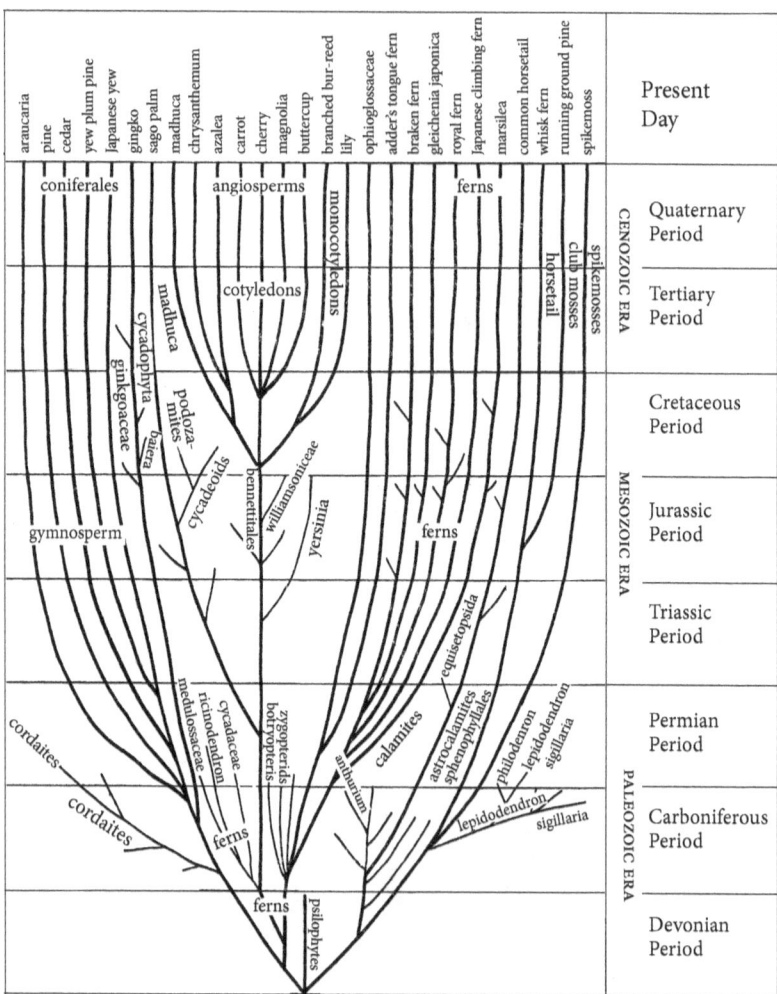

Geschlecht und Geschlechtsbestimmung im Tier- und Pflanzenreich, Max Hartman further demonstrates a clear principle of selection at work.

There were three stages in the evolution of asexual reproduction. The first stage saw the division of cells. During the second stage, particular spores, which are neither male nor female, appeared during fixed periods. All cells have this capacity, but without it algae would not be able to propagate. At the third stage, it became possible for particular cells to create spore cases or sporangia. Not all cells were able to differentiate so as to eject spores.

At a further stage of development algae became sexual as well. Sexual reproduction is also divided into three periods. In the first, aquatic plants generated egg cells with no means of protection and these zygotes, as they are called, stuck to a large stalk. The so-called gametes with small tails were male and the larger ones receiving the males were female. In the second period, a protected sac appeared along with a terrestrial mother and father. In the third period, male and female trees separated the way ginkyo trees do and, as inconvenient as it seems, sexually differentiated plants died out.

Plants like cucumbers and corn appeared in which male and female flowers were gathered into a single plant. Other plants appeared with the male and female in the same flower, as in the case of the chrysanthemum. Of course, sexuality is also active at the lowest level of kelp and spirogyra. Even tiny little spirogyra and the sea lettuce that spreads in water tanks used for fire protection have inhabited the earth already for hundreds of millions of years, having been prepared to accommodate themselves to changes in the sun, the galaxy, and the rest of the heavens.

The evolution of beauty

In studying the evolution of plants and flowers, one is immediately struck by the fact that beauty did not evolve very far in sea plants and vegetation that flourished during the early stages of the earth's history. As plants ascended to dry land and came to live in the atmosphere, pollen was carried through the mediation of insects and birds, which led to a remarkable evolution in their beauty. Darwin noted this evolution of beauty in the development of the "eye" design seen in the tail feathers of pheasants.

The pistils of wind-pollinated flowers and flower petals did not evolve aesthetically, but along with the appearance of invertebrates and vertebrates on the land, we cannot fail to note the striking patterns and colors on the petals of flowering plants. Considerable evolution took place in the color of butterflies and ants, and at the same time flowering plants evolved aesthetically—something not to be overlooked in a theory of cosmic purpose. The inescapable fact is that plants in general underwent an aesthetically pleasing transition in form, and along with that the beauty of "flowers" evolved through bisexual selection.

6 Attuning to the Struggle for Survival

Adjustment in the struggle for survival

The intake of food and the struggle for survival

Certain enzymes are produced in the bodies of lower organisms and stored there, but their production is more complicated in higher life forms. In order to collect these enzymes, the higher need to kill the lower and feed on them.

For example, protozoans—especially ciliates (infusoria) and single-celled bacteria—breed in the first stomach of mountain goats, who digest their food by passing it from a first to a second stomach. Bacteria breed by ingesting the ammonium produced within the goat's stomach, and feeding on these bacteria, ciliates grow and breed:

1. As Kandatsu Makoto and Takahashi Masami of Tokyo University have found, this is the reason vitamin B_{12} is contained in the milk of mountain goats.[1]

2. Furthermore, lower organisms accumulate chemical elements which higher organisms then absorb.

3. Vegetarian animals have the capacity to transform fiber into glycogen and turn it into sugar. Humans have a reduced capacity in comparison with mountain goats and milk cows, which is why they drink the milk of these animals.

4. Lower life forms have the ability to turn inorganic elements into organic ones. Higher life forms, which lack this ability, use the lower as

---

1. See Kandatsu Makoto 神立 誠 and Takahashi Masami 高橋正躬,「反芻胃の消化に関する研究（第四報）Infusoria の人工培養に就いて（其の三）[Studies on reticulo-rumen digestion. Part 4: On the artificial culture of some *entodinia* (III)],『日本農藝化學會誌』[Journal of the Agricultural Chemical Society of Japan] 30/2 (1956), 96–9.

Attuning to the Struggle for Survival

a food source and thus take in organic elements in order to accumulate chemical elements.

Laws governing the struggle for survival

General laws:

1. The total volume of organisms, seen from the history of the earth, has remained by and large the same from the Paleozoic period to the present day (Vernadsky).
2. Major catastrophes aside, species are largely continuous.
3. Animals depend on plants and do not exceed them in total volume.
4. Where the total volume of plants is low, that of animals is also low. Accordingly, there are few species of animals in cold regions and opportunities for the interactive struggle for survival among animals are few.
5. In primitive times the struggle for survival among animals was fierce.
6. In general, lower organisms become food for the higher. Without them, higher life forms become extinct.
7. Death is interrelated with marriage.
8. Before they die, many lower organisms are absorbed into the higher as nutrition.
9. Many higher organisms produce young in expectation of periods when plants flourish (spring and summer) and when lower organisms flourish.
10. When propagation of one species becomes too great, the balance in nature breaks down, and as a result small animals accustomed to vegetation turn to the consumption of meat.
11. There is no struggle for extinction (J. H. Fabre).
12. Even in the case of defenseless individuals, survival of the species continues.
13. Certain things like insects escape through radical physical change.
14. Plants and certain species of organisms survive by being consumed by the organisms that feed on them. (A) Higher organisms protect and nurture lower organisms within their oral cavity, stomach, appendix, and so forth. (B) This serves as a protection of the species.

(c) This is less a service than a protection (as seen in ant parasites and the like).

15. Beasts of prey can cohabit with herbivores. While paying the price of their blood as victims, they also receive protection from these beasts. Herds of zebras and wild goats are protected by lions. Silver chimaera sharks protect bonito.

16. Carnivores can be morphologically altered by herbivores. (As the history of ancient organisms in North and South America makes clear, there are saber-tooth tigers who have mutated into vegetarian camels.)

17. There are cases where, in order to protect the balance of nature, carnivores play a policing role. (In the Tertiary period of ancient Egypt, vegetarian kangaroos were transformed into carnivores to succeed beasts of prey that had disappeared.)

18. Due to stomach capacity and territorial problems related to the same species, even the carnivores among higher organisms do not eat more than a fixed amount of meat. Carnivorous beasts of prey are outnumbered by herbivores.

19. There are cases of animals who are completely unprepared to protect themselves or attack. Nevertheless, they endeavor to protect the species by means of a very strong ability to multiply (as with bacteria).

20. Physiologically speaking, lower animals cannot absorb nitrogen atoms fully into their own bodies the way higher animals do.

Parasitism:

21. In contrast to animals engaged in struggle for survival, we see animals coexisting in the same habitat. In particular, there are those parasitic on other animals or living in the same nest. (Among other things, this has been observed in ants and birds.)

22. Parasites are numerous in animals with substantial vitality.

23. Tropical plants show a greater tendency to cohabitation and parasitism more than to a struggle for survival.

24. Many insects shirk the struggle for food and show a tendency to depend on certain plants for their food.

25. Higher animals avoid the struggle for survival through collaboration or by effecting a territorial or regional pact.

Motherhood and the evolution of mutual aid:

26. Many of the lower animals lead a communal life (mosquitos, flies, locusts, crabs, frogs, and so forth).
27. Communal life often provides mutual aid in resisting enemies (Peter Kropotkin).
28. Many bird species live in groups within the same territory as other species, completely transcending the struggle for survival.
29. In higher animals protection is provided for physiological insemination and fertilization. The evolution of seed embryos in plant seeds and of placentas in mammals are examples of this.
30. Motherly love and social love has evolved among higher animals (as in ants, bees, and vertebrates.)
31. The struggle for survival by and large exhibits four stages and forms a pyramid structure (Charles Elton).
32. There is a general tendency for larger animals to feed on the smaller.
33. Organisms living in the desert mainly survive through adaptation and the struggle for survival is almost absent.
34. The struggle for survival is fierce among deep-sea fish.
35. Small animals have a simple physiological structure, giving them greater vitality in comparison with large animals. Even falling from heights does not result in death. Many of them do not die when part of their body has been eaten by an enemy.
36. Certain types of animals can live without food for fixed periods, even when making voyages that should kill them. In seeking refuge to avoid an enemy attack, they may even use an opening behind the very place where the struggle for survival is going on (eels and tarantulas).

Camouflage:

37. Insects camouflage themselves by changing shape, color, and so forth in order to elude the attack of an enemy. This may be said to constitute a physiological mutation with a finality.

Restrictions on the struggle for survival through physiological specialization:

38. Physiological restrictions on the struggle for survival are present in invertebrates and vertebrates.

39. The animal world restricts the struggle for survival by differentiation in six directions: (A) aquatic life, (B) life underground, (C) life in a shell, (D) life on the run, (E) life in the skies, and (F) life as a carnivore.
40. Birds that live in the sky enjoy the most peaceful lives. They do not burrow themselves underground or live inside of a shell. From this alone we understand how peaceful their lives are in comparison to insects. (Specialists say some eagles in Japan live one to a valley.)
41. Wild beasts that sharpen their fangs and claws lack the power of flight, while things that can fly, despite their beaks, do not have poisonous fangs. In this sense, the ability to live is restricted.
42. Certain insects possess the enzymes to digest proteins and know how to absorb the sinews of their enemies. But their proboscises cannot be used for other things. The restrictions are severe.

Ecological restrictions on the struggle for survival. The struggle for survival shows strict geographical, climactic, and biological restrictions:

43. Restrictions differ for sea and land.
44. Things in the sea are restricted by the ratio of salt.
45. There are restrictions due to the velocity of the current.
46. Even in fresh-water organisms notable restrictions are seen due to the distinction between upstream and downstream currents in estuaries, lakes, and river.
47. Differences in water pressure reflected at the surface, in middle layers, and in deep layers affect the species of organisms living there. Land animals, too, are restricted according to differences in latitude and altitude, distinctions stemming from straits, and deviations in atmospheric pressure, wind pressure, humidity or dryness, temperature, frigid zones, warm zones, and tropical zones.
48. Variations in the temperature of water and air restrict and condition the movement of fish and birds.
49. For organisms living in the ground, differences are created in habitats through the nature of the soil, temperature, humidity, and the ingredients in rocks.
50. The form that the struggle for survival takes varies in different vegetation zones.
51. Differences in the microorganic realm affect the struggle for survival

among other organisms. In the higher animals we find a tug-of-war agreement over the place where food is hunted.

Temporal restrictions on the struggle for survival:

52. The struggle for survival among living organisms is restricted by seasonal changes.
53. Species of insects feeding mainly on vegetation fluctuate according to the season, which in turn causes changes in the struggle for survival among other animals that depend on insects as their staple food.
54. Changes in the struggle for survival take place according to whether it is day or night. These are mainly due to physiological factors.
55. Even among animals living in the daylight, different strategies in the struggle for survival are adopted for early morning, midday, and dusk.
56. The form of the struggle for survival among organisms differs according to whether they are in a stage of infancy, growth, or old age.
57. Animals like birds and fish that enjoy a wide range of movement comply with changes in climate.
58. Certain types of birds (like terns) travel back and forth between the north and south poles. Flocks of starlings circle the Pacific Ocean in a counterclockwise motion. There are also types of fish that wander great distances. The migration of whales often covers a vast circuit.

Adaptation in desert plants

The fact that there is absolutely no evolutionary struggle for survival among plants is easy to understand by considering desert plants. I once visited California's Death Valley, a curious place that lies below sea level like the Dead Sea of Palestine and where the salt-water sea of former days has evaporated, leaving a plain of crystallized table salt.

Fritz W. Went, who conducted detailed research on vegetation in the neighboring region, has made some interesting observations. Because of the high concentration of salt and the paucity of water, only particular families of mesquite, creosote, peucephyllum, holly olive, and the like adorn the desert, all of them with special traits. Desert mesquite extends its roots straight down from 30 to 100 inches as a means to draw water up to the surface. The peucephyllum seems to live even without water. The desert holly olive has white leaves and is very resistant to salt.

With a modicum of rainfall, paloverde, ironwood, smoke trees, and the like spring up in an area converted for a brief spell into a riverbed. Paloverde grows like a shrub with only a few buds sprouting under the tree. Many grow at a distance from other plants. The husk of its berries are hard and it is only in times of flooding when pebbles scrape against the husks and wear them down that they can bud. As a compensation, no animals feed on them, giving many buds a chance to grow.

Many plants in desert areas show selection in this kind of budding. According to Went, the conditions are almost like those attending human childbirth.

The balance of nature

The struggle for survival cannot be considered the only cause of evolution. The struggle is real, but it is no more than a surface tension that does not touch the inner causes of evolution.

Besides the struggle for survival there is selection involved in the choice of a mate as well as psychological sifting. Even if we limit ourselves to the interface of the struggle, the law of the jungle—the strong eating the weak—does not shape the internal mechanism of evolution. The psychological predominance of ants and bees cannot be explained only by reason of their strength in the struggle for survival. There is no denying the fact that this depends on a comparative evolution of systems organized around mutual support.

First among the victims in the struggle for survival are the plants; insects follow next. But morphologically all plants are unarmed. The greater part of them are eaten just as they are, without resisting.

Insects, in comparison, are sensitive to the struggle for survival. All fish, amphibians, reptiles, birds, and mammals have vegetation as their staple food, but for many, insects come second. More than that, plants also look to insects as a source of life, as seen in the strange phenomenon of plants that capture insects (butterworts, sundew, and the like).

Higher animals must somehow take in several sorts of animal amino acids and proteins. Given the fact that these are not necessary for plants, the snagging of insects represents an irregularity. There are mysteries in the balance of nature that remain hidden to us.

From an evolutionary point of view, higher animals look to plants with the ability to create organic elements from the inorganic, and the energy produced through the consumption of these elements may be thought of as channelled into physiological and psychological evolution.

Attuning to the Struggle for Survival

Diagram of the Struggle for Food in the Mayfield Caves

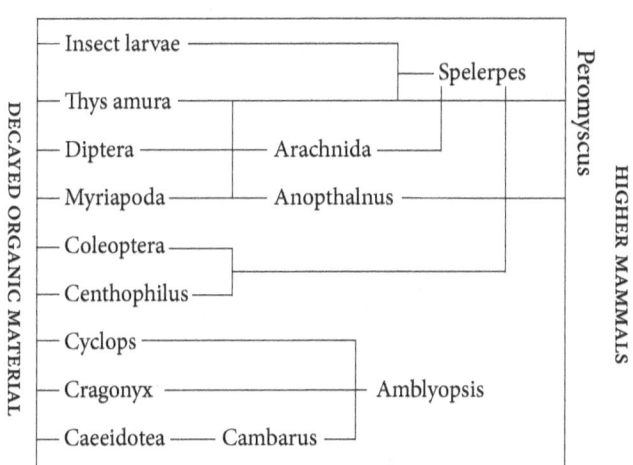

According to studies by A. M. Banta and others, killing off a species of weasels that consume the flesh of seventeen types of cave-dwelling insects had the effect of rendering these seventeen types of organisms completely extinct.

This research also helps us understand better the way in which the existence of what are apparently brutal carnivores serves to regulate the populations of several species. The diagram above illustrates the struggle for food in the Mayfield caves where Banta carried out his research. In it we see just how complex the struggle for food among living organisms is. Attention should be given to Banta's conclusion that the killing off of one animal can totally destroy seventeen organisms. That is, even a small flesh-eating weasel has a necessary function to play in the community life of the caves, and should the weasel *not* eat other animals, some species would multiply while others would be pressured to the point of extinction. This is seen on a larger scale in the appearance of carnivorous mammals.

There were no simple flesh-eating mammals in the Quaternary Period, and in the absence of carnivores at the time, many of today's marsupials with herbivore tendencies were fitted out with sharp incisors that resulted in their playing the role that carnivores play today.[2] Thus carnivorous species may be said to play a kind of policing role. Banta reported that

---

2. Henry Fairfield. F. Osborn, *The Age of Mammals in Europe, Asia, and North America* (New York: Macmillan, 1910), 79.

small herbivores, like the tree sparrow in La Plata, Argentina, went crazy when rats were around and attacked them. Obviously, certain extremes of propagation would result in something like a great cosmic conflagration which allied forces would have to safeguard against. If this is the case, it is not impossible to think of a certain species standing in reserve to carry out their mission.

This is only an image, of course, and it should not lead us to suppose that carnivores have been assigned a cruel mission. Incomprehensible though it be, we must recognize this as an important and interesting matter that surpasses the powers of human imagination. If we follow Darwin to the point of thinking that this is all there is to the evolution of organic life, we come to the rather romantic conclusion that "eating flesh is an evolutionary advance." Rather than ascent to such heights of speculation, I prefer to think that something mysterious is going on here. I even find it easy to understand the life world that Walt Disney filmed in Central Africa where lions and wild goats live together.

The Essence of Cosmic Purpose

7 Knowledge of Cosmic Purpose

Recognizing purpose: contradiction and chance

Is there actually purpose in the universe? Natural calamities follow one after the other, human life is full of suffering, and the struggle for survival is never-ending.

Infants are born in a painful delivery, only to suffer illness, eventually be visited by old age, and one day have to face the the angel of death.

Can we still speak of a purpose to human life?

The stars revolve without a care. Matter condenses without showing any emotion. Can there really be meaning in the material world? Shakyamuni the Buddha taught that all things are empty; Darwin advocated a theory of evolution on the supposition that the universe is a blind machine.

And yet, Darwin himself did not consider his keynote mechanistic view the only absolute. As a letter responding to a Swiss student confirms, he left open the possibility of rethinking evolutionary theory from the standpoint of cosmic purpose.

Why did Darwin, despite his doctrine of blind mechanism, partially affirm a teleological point of view?

How may we come to know cosmic purpose? Is there ever really a way to discover it?

To begin with, let us try to ascertain the essence of the universe on the assumption that it is completely blind, a mere chance assembly.

The impossibility of infinite chance

First of all, infinite chance is not possible. If chance were infinite, knowledge itself would not come about and *finite* assembly would be impossible. This leaves us no choice but to conclude that absolute chance means lawlessness and that there cannot be infinite chance in a finite world.

Secondly, once absolute chance has been disallowed, chance must be restricted in a finite world. Where was this "finitude" cooked up? Finitude means the determination of a "field." Both spatially and temporally, *field* and *finitude* signify the exact opposite of chance.

Thirdly, in a world whose *field* is determined, random chance is itself restricted. The more these limits are imposed, the closer we come to lawfulness and to the possibility of recognizing in chance a degree of freedom to change within the conditions placed on chance.

For example, the numerous electrons flying about among the seven energy levels of an atom appear to be circulating by chance; in fact, they have a splendid arrangement and selectivity dependent on a certain degree of freedom.

On the great authority of the atomic physicist Arnold Sommerfeld, we know that there is a design in the minute world of atomic activity.

Finally, what can it mean that the old mathematics of probability, once mustered as proof of casualism, has given way to statistical probabilities?

Even though chance should increase the larger those numbers become, in fact they are inverted into a probability with a new meaning. On what grounds is this so?

There are a few cases in which chance is visible, as in Mendel's law of probability or in the likelihood of sex and death extrapolated from population statistics, but the fact that we draw closer to probability the higher the numbers get presents a phenomenon diametrically opposed to the law of chance. What is the basis for this?

Consider a family in which a girl has just been born. We would think that the sex of the children in the family would be determined by chance, but we if take the birth statistics for 1,000 cases, we find that the number of male and female children is close to even.

If this were a matter of chance, the probability would rise as the numbers increase. For example, instead of having 500 females and 500 males in a sample of 1,000, suppose that there were only one female and 999 males; or that 999,999 out of a million children were male and only one female. In that case, we would certainly have to conclude that chance was dominant. But what may not hold for a group as small as 100, when we enlarge the statistical sample to 1,000, independent of time and place and in the absence of special reasons, the numbers would even out. Raise the figures to 10,000 or 100,000 or even 100 million and the ratio would still hover around half male and half female. Whatever the reasons for this might be, they are not under the control of chance.

Those who do not think very deeply about these matters claim that

Knowledge of Cosmic Purpose

Mendel's genetic law is the result of chance. That law unfolds in a quadratic equation:

$$AA + 2AB + BB$$
$$\therefore A^2 + 2AB + B^2$$
$$\therefore (A + B)^2$$

Some claim that inheritance takes place by chance. But if that were so, we would expect random results that did not necessarily yield a quadratic equation, namely:

The elements AB do not overlap regularly in (1) but join as AABB, isolate B to make AAB, or conversely isolate A to make BBA, or again eliminate AB altogether to combine as AA, BB, or conversely duplicate AB so that we would not expect the combination AA and BB. That AA, BB, and AB nevertheless show up is proof that chance is not in control.

In this way, because the joining occurs perfectly in the union of A and B with their A and B counterparts, Mendel's law is not chance and a certain regularity has to be acknowledged. Particularly surprising here is the doubling that shows up when the union AB joins to the other AB. They do not work as a new unit but split into two, each of which works separately as do each of their counterparts. Could something like this occur in a chance world? Is it not altogether unthinkable that such a thing could happen without some prearranged power to unite? Is this *power to unite* the work of chance? Is there not finality contained in the *uniting* itself? To think of this uniting in terms of chance, all arrangements would have to occur by chance. To speak of "arrangement by chance" is to quibble over words that mean the same thing. In the case of small numbers, we ultimately come to realize that what appeared to be a matter of chance was actually something constructed with a design.

The same holds true in calculating death statistics. Say someone has died. Family members may all think that the death just happened. But viewed statistically, it is nothing of the sort. The death could be understood as a necessary outcome associated with conditions in the individual's environment, physiology, psychology, and morality.

The refraction of purpose

There is we say a kind of *progressive* quality or weight attached to purpose. In the case of a machine, some of the parts in the assembly may be thought to serve the purpose of conserving energy. In addition to the overall change entailed in the assembly, we can think in terms of the effective energy of the machine to do work.

In speaking of energy, there is something like effectiveness through selection, and in the case of growth, there is an increased capacity for growth. Both are necessarily accompanied by progression towards a goal.

Since possibilities for weighting are always present by arrangement, it follows that purpose cannot be simple. This is precisely why purposiveness seems relative and, like energy, cannot be grasped simply and intuitively.

The structure of purpose

The elements making up purpose

There is arrangement and selection in purpose, and there is also the need for a drive toward purpose. Further, there must be strength to sustain a purpose. As a result, there is a also a need to determine the orientation toward a purpose.

Prearranged purpose means lawfulness; purposive selection means sifting; advance towards a purpose means growth; orientation to a purpose and continuity in the strength to pursue it mean restriction of the mechanical structure and the field. In other words, a changing world entails the utmost in restriction. Put this way, we notice a structure present in purpose by arrangement.

In order to realize a purpose, simple though it might be, five elements are necessary: energy, change, growth, selection, and law. These five are complementary, but it is energy that first sets the pursuit of purpose on its way. Energy accepts change, grows into purpose, eliminates all sorts of obstacles, chooses the best path and method for the purpose, preserves adjustment to the surrounding conditions and environment, and protects the various arrangements and conditions that allow it to ascend to its goal. These five factors together manifest the *power of life* within the living world. In the conscious world, the addition of this power of life as a sixth element lays the cornerstone for self-determining purpose.

There is nothing very complex about the structure or organization involved in setting up a target at an archery range. This might lead us to think that nothing more than ordinary commonsense is required to establish a purpose.

In fact, purpose is a configuration built up of five requirements: (1)

progression, (2) mechanism, (3) law, (4) mobility, and (5) selection. Even in the case of purpose in the inorganic realm, these five have to be taken into consideration. When it comes to the organic realm, life is added as a sixth requirement, and the advance to self-determination requires an additional mental or conscious element. If this were not the case, it would be completely impossible to suppose a purpose, let alone realize it.

The relativity and progression of purpose

Purposiveness is relative. This means that it is

1. oriented: A turns in the direction of B;
2. adjusted: A adjusts to B;
3. progressive: A relays the execution of purpose to B;
4. discriminating: A is distinguished from B;
5. arranged according to the law of cause and effect: A becomes B;
6. instinctual: A desires B; and
7. conscious: A has B as its ideal.

The opposition of A and B in the above is relative and has an element of progression.

In traits 3, 5, and 7, this progression is seen in the turn to the future present in the arrangement. Ontologically speaking, that future does not yet exist. Thus a systematic mechanism is needed to link A and B and to set off in pursuit of a world that does not yet exist. Further, traits 1, 2, 3, 4, and 6 require a mechanism for keeping A and B connected (see the section on the epistemological existence of purpose later in this chapter).

The nature of the relative traits (1, 2, 4, and 6) differs from one to the other, but they all preserve an intimate connection between A and B, and expect the exercise of selection under fixed conditions subject to the temporal and spatial restrictions of the "field." In this sense, there is no way for the determination of the field and a fixed movement to achieve a purpose except through mechanical operations.

Even the simple case of a stamen adjusting to a pistil entails the conditioning of a mechanical apparatus. The same can be said of an electron selecting an orbit within an atom. Ordinary electrons allow flight from the contour towards the interior, but flight in the opposite direction is not permitted without the addition of special energy from outside. There is a determination of direction here.

This also holds in the case of the instinctive drive for food. While there is no mechanical apparatus in the food as such, if it is absent in the organism, the goal of eating could not be achieved. It is not by chance that a parasite settles in an animal. From the beginning it must set its sights on the right time to act. Moreover, in order to settle in as a parasite, an adequate mechanical apparatus to that end must be in place. The principal point in the structure of purpose here is the requirement of a mechanical operation involving selection to make the connection between A and B.

The need for the mechanical in purpose

The future does not yet exist; it is nothing. To carry life into this nothingness is an attempt to transfer power or energy. In other words, it is the attempt to orient a certain movement and give it continuity (which already includes a spatial and temporal purposiveness). The purpose is to transport *life* and *energy*. Only a mechanical construct can achieve this.

Transporting energy into a realm of nothingness and giving it continuity requires the determination of a *field* capable of prevailing over all sorts of variations. The determination of this field must provide a continuity that limits change and progressively transfers the *being* of energy into a realm of *nothingness*. For this to happen, every part in the assembly has to be spatially unified and temporally controlled so that, beginning with an orderly progression, *energy* can be carried in a fixed direction and a selective determination with fixed rules can be included in every form of change. Only then can it achieve the purpose of providing continuity in energy or life.

This unity of space and control of time cannot come about without selection. Mechanical assembly cannot appear as a sudden event relying on chance or blind fate. It requires selection within change, an orderly determination of direction that permeates the progression of temporal change. Because this determination of direction presupposes temporal change, it is not predestined to lack selection. We may suppose it is endowed with a mechanism as the condition for variation, progressive growth, and selectivity.

Continuity in *energy* and *life*, no matter what else, needs mechanical facilitation. It has been thought that machines do not have a purpose, that they are purposeless entities, things that "work like machines." In fact, no machine is without a purpose.

That said, this *purpose* is not something created by the machine itself; it is given from without. At times, of course, machines do not actually serve their purpose. One may imagine a machine with a broken part that leaves it spinning uselessly. While this sort of activity can be thought to take place in the universe, from a cosmic perspective, it is not that there was never any purpose but that it had one and lost it. Just because a certain machine breaks down and starts whirling idly, we cannot conclude that all machines were created merely to break down and whirl idly.

Our reason for supposing a certain degree of intelligence in its make-up is not just the fact that it is a machine but that it can also end up whirling idly. Otherwise it would make no sense to speak of "whirling idly." Such activity assumes another side to the picture, namely that the machine can whirl in a purposive way.

Hence, in turning to the future in the attempt to realize a purpose, the fulfillment of a purpose is impossible without taking into account some means of mechanical assembly. This leads us to the thesis, "no purpose without a machine," and to its converse "no machine without a purpose."

No purpose without a machine

Machines operate. They revolve continuously and merely go round in circles. People who observe them may think of machines as aimlessly spinning, blind entities. Now if continuous operation is not a matter of chance, neither is it simple change. Here order, law, and selectivity are hidden.

We need also to recognize latent purpose here. Earlier I explained that because purpose is detached from the present, its achievement requires the working of a mechanism.

However, an "entity" is the product of a past with the capability of giving birth to the present. For this to happen, there must be a mechanism with a purpose for giving birth.

Existence comes about through a chain of purposes. Even if an entity does not have a final purpose, purpose can be given to it in the form of an intermediate purpose or as the result of a purposive progression. Entities always have a powerful mechanism behind them. Even when the existence of a machine is not realized in some final purpose, it was clearly made for some intermediate end. The presence of the phenomenon of continuity in the form of a single machine "in operation" cannot be the result of chance succession. Repetition of the same event cannot be attributed to chance. It has an order, and to find this order preserved in the midst of change leads us to see a kind of selection at work within change.

A spinning wheel is spinning thread and the spindle flies off, leaving the wheel whirling about idly. May we then speak of its existence as blind? The spinning wheel was fashioned by a craftsman with a fine head on his shoulders. It is a construction with a fixed purpose. What does it matter if one part flies off so it can no longer fulfill its purpose? We cannot look at the spinning wheel whirling idly and deny that there is an order to its completed form and that selection is involved in assembling its various components.

Purpose comes in more than one form. The spinning wheel does not have the self-determining purpose of the human being, but insofar as it was constructed in order to realize the human purpose of spinning thread, it has a kind of intermediary purpose. On the analogy of human activity, it "achieves its purpose" by spinning thread. Even so, the finished product is not itself the purpose. The ultimate purpose for which it was made is to keep the human body warm. Purpose comes in various types and gradations, but the fact is, no machine comes into being without some purpose.

The mobility of purpose

When disassembled and analyzed, purpose becomes incomprehensible. Purpose is synthetic and only becomes clear in its actual assembly.

Similarly, when the movement of something is halted, its purpose is completely lost on us. As stated earlier (1) A moves towards B, (2) A is linked to B, (3) A transmits energy progressively to B, (4) A is distinguished from B, (5) A becomes B, (6) A wants B, and (7) A has B as its ideal. If all these conditions are not met, the goal cannot be achieved. This shows the mobility of purpose.

Dead things have no purpose

When things come to a halt, there is no knowing where their purpose lies. Anatomists only dissect dead things. The reason they cannot see the finality of living organisms is that they do not grasp the purpose of the mobility that belongs to life. Only by cultivating blood for long periods and looking at how white corpuscles divided to create blood vessels for the circulatory system was Alexis Carrel able to discover the finality in blood. So too, only by allowing a machine to work can we discover its purpose.

The lawfulness of purpose

Purpose is not to be found in disorder. To say that purpose can be realized through mechanical mobility, means that there is a lawfulness to purpose. The purpose realized when a seed germinates, blossoms into flower, and produces fruit, obviously includes a law of cause and effect immanent in the sequence. Even in the absence of the law of cause and effect, the laws of dynamics governing machines may be said to bestow their design on the inorganic world.

Even when the physical laws of mechanics are not in force, the laws of chemical coordination are immanent. Such chemical adaptation shapes the principles of chemistry, and this adaptability in the inorganic realm ultimately forms the basis for the impulse to life in the organic realm.

In the struggle for survival as well, the marvelous and robust indomitability of statistical laws leaves us altogether astonished.

Simply put, there is a need for order and organization in the realization of purpose. When an arrow is aimed at a target, first an arrow is nocked and a direction set, then power is exerted to pull the bowstring, and finally the arrow is sent flying towards the target. A mechanism is needed to realize the purpose and this is why laws have to be provided.

The selectivity of purpose

When determining direction based on coordinates, choices have to be made. Shall I go to the right or to the left, above or below?

Such determinations are never a matter of chance. The organic chemist Wolfgang Ostwald observed a tendency in atoms and atomic compounds, which he referred to as "directionality" rather than purpose, but we should also note the selection involved in the determination of direction.

This may not seem like much, but it matters a great deal. The development of higher organisms, like human beings capable of self-conscious decisions for a purpose, is also grounded completely in the tendencies of atoms. Were atoms not endowed with design, cells would lack the capacity to adapt and to organize the nervous system and sense organs. If we think of a high-order finality coordinating and bringing about lower levels of finality, there is great significance to Ostwald's treatment of the directionality of atoms as the first steps on the road to "value."

The accommodation of light to the eye, sound waves to the ear, tastes to the tongue, and smells to the olfactory organs involves an extraordinarily intricate mechanical apparatus. Selection sees to it that sound waves do not enter the eyes and light waves are not reflected in the ears. There is clear selection seen in the limits imposed on the length of the light rays that can be reflected in the eye, keeping infra-red and ultraviolet rays out.

In a relay race, strict selection insures progression towards a goal. So, too, the rigorous regulations of organic chemistry have to be followed in order to maintain the progressive stages of the digestive system. Only when the duty of relating the purpose from one stage to the next is carried out can the final stage be reached.

When he was a young man, my friend Takeuchi Masaru, who worked in Kobe for many years as regional director of the Ministry of Health and Welfare, heard me speak about progression in the digestive system. Coming to appreciate the need for mastication, he started to chew his food thoroughly and never to put hot water in his rice. I find it interesting that even today when fish bones and other small bits of foreign matter find their way into his stomach by mistake, within an hour or an hour and a half he always experiences a reflex of the undigested material from his stomach to his mouth.

In considering how each of the various physiological organs fulfill their own progressive and relaying purposes, once again we are forced to acknowledge processes of selection that surpass our imagination.

A RESOLUTION OF KANT'S ANTINOMIES

Examining the Kantian antinomies

Why did Immanuel Kant propose his antinomies? Looking at the world of nature, he argued that it is not necessarily impossible to take a finalistic view of the cosmos.

In hindsight, Kant lived in an age when the materialism of Pierre-Simon Laplace was in full bloom and the French encyclopedists wielded enormous influence. Obviously, such thinkers assumed the antinomy of matter and spirit, of mechanism and free will, of chance and purpose.

When we come to the study of matter, the differences between Kant's age and our own are immense. In his day, matter was held to be bound exclusively to the law of cause and effect; it was subject to chance, without order and without selection. Unlike the wave mechanics and quantum mechanics of our day, which tell us of selectivity and the freedom of the uncertainty principle, they knew of nothing that (microscopically) transcended the law of cause and effect.

Despite the finality involved in Kant's *Critique of Practical Reason*, at the time the natural world could only be conceived of in terms of opposites, thus generating a cosmology of antinomies.

Even today, those unfamiliar with physics probably consider those antinomies to have been displaced by common sense, impossible as it is to revert to a pre-Copernican age when the sun was thought to revolve about the earth. If such opposition did exist between the material and spiritual realms, there would be no reason for that special something we call spirit to be generated within the material flesh. (If the term *generation* is problematic, we may also use *emergence* or *concurrence*.)

Common sense tells us that we are nourished by matter and engage in a spiritual life without any opposition or contradiction. One school of thought avoids antinomy but posits the interconnection of two opposite and incompatible realms: the subjective and the objective. But to relate an objective world of matter tied to the law of cause and effect, with a subjective world ruled by free will, in such a way that never the twain shall meet, is to end up in just another form of antinomy.

The time has come for the antinomy to be sublated. As we see in the interaction of the body with mental functions, there is a purposive design built into the essence of matter and arranged from the beginning.

Antinomy, the relative world, and temporal transcendence

Ordinarily, it is easy to think of a relative world in which purpose and blindness stand in relative opposition, but this does not imply logically that a relative world is also necessarily blind.

First of all, a relative world seems to exist as a confrontation of two or more opposing forces. From the perspective of "relationship," opposition is a form of relationship but not a "separation." At a different time and place, the wind can seem to be moving in the opposite direction. Because the wind goes counterclockwise in the northern hemisphere, it blows at first from the east and finally from the west. Thus what appears to be an opposition, when viewed calmly, turns out to be an identical movement in a single system.

If we think of opposition and antinomy in terms of plus and minus within a frame of relationship, what appears to be opposition and antinomy, when seen from the perspective of the whole, turns out to constitute a continuous relationality. Day and night occur as the earth spins; the plus and minus of male and female are an expedient opposition for reproduction. Or again, consider the plus and minus of electricity. From the viewpoint of wave mechanics, when the upward and downward movements of a continuous wave are obstructed at a given point, both movements appear to be opposed to one another. But when the minus electron and the plus electron that make up opposition are joined together, they are transformed splendidly into "light," as the new quantum mechanics has proved.

Put this way, antinomy does not signal an absence of relationship but relationship in an advancing world. When we examine this relationship, we find that it is not without purpose, and indeed would be meaningless if it were.

Relationship *depends* on certain conditions. This dependency includes its being *for the sake of* something, and it is there that purposefulness lies.

Opposition between old and new ideas occurs *for the sake* of some situation or *depending* on something or other. Even apparently blind, chance events, if looked at closely, have deep causes. Why could a certain Indian philosopher who emphasized "emptiness" not sense any contradiction in adhering at the same time to the law of cause and effect? If the law of cause and effect is made central, "emptiness" is a logical impossibility, whereas if "emptiness" is made central, the law of cause and effect does not arise. This is because a certain type of Indian philosophy which integrates these two poles without seeing them as contradictory has transcended antinomy.

A world that can transcend is a unified world, not a world of antinomy. Antinomy can be transcended by changing levels. So it is with Kant, who held that the antinomies of knowledge can be transcended in the realm of free will. He thought of *knowledge* and *will* as two distinct realms. But actually, in the sphere of *consciousness* they form a single unified whole capable of developing into a continuous realm of purpose.

"Time" flows in one direction. If the realm of antinomies runs parallel on that same temporal axis, advancing in the same step-by-step transcendent motion, it would also be dragged into the flow of a unified world.

Reversal is disallowed in the world of time. In other words, time is linear and directed towards a goal. At a time when the future looks very dark, it can happen that one looks back on the brilliant record that has been left behind to see an "evolutionary" tendency hard to dismiss as blind.

The probability of time being oriented to a goal requires too broad a vision for it to be perceptible from what lies immediately in front of us. But if, for example, we look at the chart of human body temperatures and compare them with past records, we are able to discern a direction.

All sorts of possibility and antinomy are involved here, but the point is that the past is an index to where the future is headed.

Finality in relationship

A relative world means a world of relationship. And all relative relationships are bound to absolute laws.

Now all law—whether in physical entities or in mental values—depends on *energy*. By prearrangement, energy has a quality of spatial change and temporal growth or progression. It is "arranged" that selection is present not only in spiritual values but in everything, right down to the last atom of physical existence.

It is, therefore, odd to hold that there is no purpose to the complex structure of this vast natural world, in which normativity in energy, change, growth, and choice is given together with laws.

The structural nature of these things is a preparation for the emergence of purposiveness, so that to exclude purpose from thought is also to exclude structure. Structure comes about precisely because there is purpose. All matter and all mental phenomena depend on energy. Energy gives value to space; conversely, space possesses "static energy." With the transition to movement, we have "dynamic energy" and, conversely, a "temporal quantum."

Hence there is not a single entity in either the material or the mental world that does not rely on energy. The change and growth in which this reliance is expressed constitute an unfolding from given causes to given effects. To revert to our earlier terms, the world of *from* and *to* possesses direction and structure.

Once energy has entered into the substantially different world of law, it cannot be expected to act without some kind of purpose. The reason is that energy, whether studied from the viewpoint of statics or of dynamics and force, never works by chance. Magnetism is made to work perpendicularly to the flow of an electric current. Selection in the natural world is not restricted to the realm of energy. There are restrictions and arrangements in change as well; and, as the Russian scientist Oparin attests in the field of catalytic chemistry, change displays a remarkable finality. In growth, the arrangement includes the finality present in Mendel's law and in the genetic factors discovered by Thomas Hunt Morgan. In this light, relationships of *depending on* signify relationships *for the sake of*, so that all of the relative world, in one form or the other, includes the makings of a constructive purposiveness.

The antinomy within purpose

If this is so, why do antinomies occur in a purposive world? The answer is that a cosmos with purpose requires it.

On one hand, in its progress through time, purpose exists *ahead in the unseen* while only mechanical phenomena are visibly present. Antinomies show up in contradictory states like those marked by the opposition between spirit and flesh, ideal and reality, goal and mechanism, thought and matter, God and humanity, infinite and finite. But if we think about it carefully, we see that the phenomenon of antinomy itself only appears where there is purpose.

On the other hand, from a spatial perspective, purpose can never be realized by simple means. A system has to be constructed before principles can be established to preserve life for a thousand years, two thousand years, or an almost infinite future. This preservation (of matter, energy, and life) cannot achieve its purpose if it is completely cut off from assembled mechanisms. The introduction of the mechanical into the realm of purpose may once again appear as an antinomy of contradictories. But here again, we need to recognize that the phenomenon of antinomy itself is a consequence of a purposiveness world.

The category of space and the relationship of finality

How are categories like time, space, choice, and law interrelated? What can they mean? What kind of design could be lurking in the shadow of these categories? Up until now, science, philosophy, and religion have tended to consider these categories as distinct units and not to give much thought to their mutual constitution.

When the category of purpose is added, the categories of time, space, choice, and law come to life. For this to happen and for purpose to become a reality, necessary measures have to be taken and plans worked out. Preparations have to be made for various changes. Change involves "expansion" and these internal relationships of expansion are reflected in the mental realm as the category of *space*.

Next, the realization of purpose progresses from change to the next stage, in which the "expansion" of time is projected into the subjective mental realm as the category of *time*. This progressive development—or growth—represents a new approach to purpose, but selection is necessary for change and development to respond to change and move to the next stage.

Selection provides an approximation to purpose, which is not the acquisition of purpose as such but a way to approach purpose one step at a time. If these efforts do not secure continuity in the next stage, the process has to begin anew. This is made possible by genetic elements that maintain the species and guarantee the laws of redevelopment.

Such considerations lead us to think that the categories of change, space, growth (time), selection (sifting), and law (order) are not so much separate and independent entities as raw materials out of which a purposive world is constructed.

In this sense, I understand categories to be interconnected and to form a developmental gradation directed towards a goal. The first step towards the goal is change or space. It takes time for the growth of a horizontal change to expand toward a goal by vertical expansion.

But it is difficult to come to purpose only with change and flux. This is where the subjective endowment of selection restricts the flux of change by the addition of choice, thus drawing still closer to the goal.

Darwin's evolutionary theory stressed blind natural selection but did so with a high degree of optimism stemming from a blind expectation that the cosmos is gradually approaching perfection. Darwinism eliminates purpose to focus exclusively on the *flow,* and at the same time supposes an unending progress from the simple to the complex, from the ugly to the beautiful, from the weak to the strong.

However, when we come to higher order change and what follows it, the selective limitation involved in "sifting" can no longer be blind. If we look only at the change as such, we might speak of it as blind or matter-of-chance. But this runs counter to reason because there is selection going on within the confines of this seemingly chance change, and because the actual things chosen to change are able to carry over (unchanged) to the next stage in an orderly fashion. By saying selection operates in the realm of accident and invariance is guaranteed by variation, we see a logical contradiction at the roots of Darwin's idea of natural selection by sifting.

Sifting already assumes an approximation to a goal. Nature can be vague, rough, and hard to get a handle on, as when it automatically assigns a particular abnormality to a given species. But it never grows wild without having some limited benefit.

If we wish then to understand the sifting that goes on in nature as it activates the four categories of change, growth, selection, and law, we are hard pressed to explain the phenomenon of life if all we have are theories of blind mechanism.

Chance is limited by the act of sifting, and this limitation carries over to the next generation in the form of a species with a certain degree of stability. This stability, in turn, allows for the formation of laws which, by transcending change, enable the repeated recurrence from species to life forms. I see this as a miracle of logic.

We can sidestep all of this and explain Darwinism's contradiction more simply by accepting the finality recent biology has discovered in the genes. There we find clear affirmation of finality in the developmental mechanisms of change, growth, selection, and law.

Adaptability and finality in evolution

In the case of things that evolve on earth by extension, it is difficult for the species to maintain continuity without acquiring change in its genes. The ability of genes to insure continuity is contained in the genes themselves. This genetic potentiality is a mystery of evolution witnessed in the environment and related developments on the earth's surface.

No matter how quickly genes evolve, when the environment obstructs their evolution, development is rejected in the species. From de Vries's study of the white evening primrose, we see that when a new species appears, it does not so much serve to preserve the species as to display mutational divergence. Seeing such a phenomenon, we must conclude that evolution cannot itself be the result of chance.

The continuity of species comes through genes and the correlative correspondence of the environment, similar to the way sexual differentiation functions in the preservation of the species. If we liken genes to sperm, the environment may be thought of as a fertilized egg. While the correspondence between sperm and egg is clear, that between genes and the environment is less clear and runs the risk of being passed over as no more than blind happenstance.

Constraints on chance

Numerical constraints. H. G. Wells made chance almighty in his *Science of Life*, but numbers bring constraints of their own. Differences appear in the vitality of atoms and molecules depending on whether or not they are lined up in even numbers or odd. When carbon elements number fewer than ten, they do not form a solid; when more than ten, they do. All of this represents a constraint on chance.

Coordination constraints. In studying organic chemistry, we learn that chemical compounds are constrained by their coordination. We cannot forget, either, the fixed design of the coordination expressed in cis- and trans-, ortho-, meso-, para-, and so on.

Constraints in spiral forms. Cosmic magnetism is generated by spiral nebulae among the heavenly bodies, and the fact that the solar system floats in the great void may not be due to chance influences.

Constraints in electrical circuitry. The direction in which magnetism is generated changes depending on the circuit. This is a good example of how chance is constrained.

Constraints on electrical generators. Differences arise in the electrical energy produced in electrical generators according to trigonometric principles. Insufficiencies in electrical energy are supplemented by condensers.

From only these few examples we can see that chance is constrained by the laws of the universe, numerical quantity, angle, shape, coordination, and other such very simple phenomena. To ignore them and assume that laws are born by chance is unreasonable, not to mention the fact that there is no way to imagine chance having the power of selection.

Chance and purpose: the rise of cosmic evil

Constraints on chance: chance and law

Infinite chance is not possible in a finite world. Absolute chance means lawlessness and the impossibility of even recognizing chance.

To say that a realm is *finite* is to say that it is determined by a *field* that sets ultimate limits to change.

The presence of chance in change is related to temporal formation and may be contrasted with finality. For example, when a baby is born with an unusual deformity, or when a locomotive runs off its normal rails, people speak of something that has happened by chance.

Chance does exist. In a world with limitations, chance is surely present wherever temporal flux is allowed. But it is a mistake to conclude that just because chance exists, there is no law and no selection, and that the universe is completely blind. Darwin recognized both a sifting within chance and the role played by law in the sifting, but in his rush to explain fluidity in the origin of "species," he did not think deeply enough about the philosophical implications of chance and law, of chance and selection (sifting). The result was a great deal of confusion in the philosophy of biology during the second half of the nineteenth century.

Altogether unable to explain everything as blind, chance sifting, he first corrected his ideas of natural selection by referring to the choice of mates among males and females. Later, through his study of worms and the like, he came to acknowledge a clear finality in physiological mechanism.

Entities in a finite world can in no sense exist merely "by chance." At the root of existence are the five great principles of law, energy, choice, growth, and change where finality is always present in the arrangement. These five great categories enveloping chance are very real, and there is a deep mystery hidden within chance that keeps it from being only chance.

Chance and sifting

Darwin held that nature does its sifting blindly. In declaring a "law of the jungle" within the struggle for survival, he forgot that sifting in nature is *not* blind and that there is purpose in the persistence of subjective life.

In other words, not *everything* in the struggle for survival is blind. The claim that the preservation of life is absolute implies that it encompasses an absolute purpose right from the start. Sifting takes place as selection adjusted to the goal of "survival."

Selection appears when the limits of a finite world are reduced further, and we have law when there is no more than a single determination given. Negotiations between chance and other categories developed as a statistical science, with the result that the older casualism has evolved into probability theory today.

Wave motion as the limits of chance

The reason things previously treated in terms of chance have shifted over to probability theory is that statistical tendencies point to a certain probability within a "field." This has been shown in statistical research on the positioning of electrons within the atom, as well as in the study of selectivity in the energy levels of electrons. This has given us to understand that in wave mechanics, the wave represents an outer limit to chance, so that, except in special cases, electrons cannot fly outside of that limit.

Even a modicum of attention to statistics make the wave-like constrictions placed on chance easy to understand.

Mendel's law in genetics reveals chance as a probability that can be expressed in algebraic formulas. It is the same with statistics on the death of living organisms and statistics on the occurrence of natural disasters.

Therefore, given the limitations of a field itself, chance occurrences within the field cannot be altogether blind.

Chance and the degree of special freedom

In physics, where variation is allowed, we speak of a "degree of freedom." This means that, according to the degree of freedom in the natural world, chance expands and creative potential increases through change.

The new quantum mechanics reports that when light waves, which have a wave motion, collide with matter, they create a new pair of plus and minus electrons. This phenomenon is probably not the result of chance. It may be better to speak of certain conditions that allow for a creative variation. This takes place as the degree of freedom increases. The creation of matter that would have been unthinkable to the natural science of the nineteenth century has come to be accepted as a matter of course.

Given that the entire physical world was thought to be determined, the idea that light could become matter was formerly taken to be an absolute impossibility. In the meantime, we have learned that light possesses a degree of freedom and can be changed into electrons. The more this degree of freedom expands, the more we are obliged to recognize this apparently chance phenomenon. And the more this phenomenon takes place, the more we may think about the creativity in the universe to turn light into matter.

Chance necessity and the shift from casualism to probability theory

With an increase in the degree of freedom, it becomes impossible to see any other probability in chance than that of statistical tendency, but probability indicates necessity. Hence, we settle on the contradictory term "chance necessity."

As we have already mentioned, for the individual, death is a chance occurrence, yet statistically it implies a certain necessity. In other words, if environmental determinants and forms of individual life are approached in terms of mathematical function, death statistics have a certain probability, but among the huge number of those who have died, any given individual death is taken as a matter of chance.

The same thing is clear in examples of birth statistics. By and large, the number of human births divide evenly among males and females. Statistics gathered for around 1,000 people in a given locale break down into 500 women and 500 men. If the standard is raised to a million, the resulting balance is still better.

Naturally differences in social environment yield minor differences, but in most cases the phenomenon of a 50–50 ratio is maintained.

The birth of an individual male child seems to be chance, but statistically it is a social necessity with a kind of purpose and never mere chance.

Computing chance through the a priori law of probabilities

In computing chance statistically, there is a fixed limit to chance displayed in the motion of waves. If we portray chance as ripples on a flat surface, we must conclude that it has an important role to play in a the realm of discontinuity.

Changes in the individual elements that go into the make-up of the universe—energy, change, growth, selection, law, life, and so forth—have a certain "leeway" in their degree of freedom with regard to space and time, and this appears to be subject to chance.

It is difficult to discover chance where laws are in effect; laws give probability to chance.

When it comes to the realm of energy, absolute chance is an impossibility. The direction taken by electricity and magnetism is not a matter of chance. Magnetic energy works perpendicularly to the electrical field. There is selectivity involved in the path an electron takes, and despite the presence of a quantum weight that often shows a degree of freedom approaching chance, a fixed transcendent law underlies the limits of possibility.

Similarly, electrons arising from wave motion can only be generated on condition that they are stationary. This condition is specifically determined with a quantum number of 4. According to Pauli's exclusion principle, the stationary condition set by the quantum number 4 allows for the presence of only one electron. Here we see the well-ordered structure of the electron orbitals.

The point at which we find an electron is identical for all electrons. Even if their positions are exchanged, there is no change at all in the point itself. The Fermi-Dirac statistical theory of electron systems is specified in accord with this notion.[1]

Atomic bonds, of course, are not chance events; they entail selection.

While chance and change have a certain family resemblance, they are fundamentally different. In the realms of physical and chemical change, we find laws of fixed selection that do not allow for the intrusion of chance. Selection regulates chance and puts constraints on it. Darwin had no trouble affirming that chance and selection coexist in the same world, but this is rather strange from a philosophical standpoint. Suppose someone holds the selection of an electron's trajectory to be a chance event. Does not selection imply that chance is impossible?

Darwin thought that sifting happens by chance, but in making life itself absolute, did he not land himself in an awful contradiction by excluding from his theory the fact that sifting takes place with the purpose of preserving life? When it comes to selection there is no such thing as absolute chance. Of course, there is no objection to confining chance to selection in the context of variation, that is, to seeing selection as a variation in the sense that it makes chance possible.

In growth, conditioned chance appears in the form of mutation. This chance is of two types: it can take the form of a histogen that can be repeated several times; or it can take a one-time, unrepeatable, and completely fixed form that is carried genetically.

Even here, what appears to be a chance phenomenon, on careful examination of its causes turns out to have arisen within a clearly conditioned context. From this perspective, Sommerfeld's studies in atomic physics show that the intensity of light generated by electrons in the spectrochemical realm is not a chance event but appears in integral ratios. This means that there is an a priori probability governing the world of the atom and the guaranteed presence of a priori arrangements in those very areas that had seemed to be under the control of chance.[2] Sommerfeld further concluded that the universe is not conceivable through the law of cause and effect but possesses principles similar to human thought that require a teleological approach.

In *The Structure of Line Spectra*, Linus Pauling also draws attention to this a priori probability. That is, in the realm of atomic physics, by pursu-

---

1. Sugiura Yoshikatsu 杉浦義勝,『原子核物理学――実験』[Experiments in nuclear physics] (Tokyo: Kyōritsusha, 1940), 1.
2. *Three Lectures on Atomic Physics*, 28. See note 31 on page 102 above.

ing the study of chance, apparently chance events are shown to harbor a priori conditions that cannot be attributed to chance. This can only be called an amazing phenomenon.

Creativity and the locus of chance

In general, chance is not easy to discern when growth is still in a primitive form such as we find in the inorganic world. Many phenomena that appear to be chance have been produced under conditions of the law of cause and effect. Even in the flight of electrons, another apparently chance occurrence, Sommerfeld and Pauling have drawn our attention to controlling "a priori conditions."

However, I do not agree with those who say there is no such thing as "chance." In the realm of organic life, cells show some degree of physical, chemical, and biological freedom as they take shape through high-dimensional division and organization. Not only do we find purposiveness developed to a high degree there, but we see multiple instances of digression from purpose resulting in various sorts of disease. *This* we may call chance.

Insofar as such chance emerges from a norm of finality, it does not represent a total rejection of finality as such. Where there is growth and evolution, life has finality as its fundamental principle. But life cannot emerge unless the five dimensions of finality—energy, change, growth, selection, and law—are in full accord.

Things become still more complicated and aggravated in the life world. There, a change in even one dimension causes an aberration and, in terms of purpose, generates events that are projected as chance.

Instead of the normal birth one would expect, children are sometimes born with deformities as a result of damage incurred in the womb. Such "accidents" contrast with a purposive world.

The universe flows. It is a cosmos in the making, and because of that, evolution takes place. The process of evolution has tendencies which in turn suppose purpose. For evolution to have opportunities to advance, it not only requires mutation but also chance resulting from a break in continuity. This is what makes possible both the appearance of genius and the birth of a deformed child.

That the world is constructed in a discontinuous fashion and yet integrated into a continuity through various kinds of laws points to a finality of purpose beyond the power of human imagination to grasp.

The universe's impulse to life that reaches for the infinite by way of the finite cannot evolve without an eye to opportunities for variation. Consequently, life leaves its footprints on the chances that accompany each such opportunity. Naturally, looked at in terms of its parts, there are things that do not degenerate. But from the vantage point of life as a whole, the development from protozoa to the human race may be spoken of as a creative evolutionary effort. When viewed from the aspect of the creation of life, not even great mutations that approach chance are cause for despair. In the end, the pace of evolution gives us to understand that the "a priori probabilities" that govern electrons also hold sway in the realm of life to which high-level chance has been attached.

Chance and the uncertainty principle

Change does not happen by chance. It comprises any number of ordered elements. All laws are related to change, so that from the perspective of law, there cannot be chance in change.

In the microscopic realm, however, Werner Heisenberg of Leipzig University's declaration of an "uncertainty principle" set up a collision with an uncertainty that eludes human knowledge by pointing to a realm that is, more than any other, beyond the reach of the law of cause and effect. That chance is involved here cannot be ruled out from the start, but as I have noted earlier, spectrochemical research has shown that microscopic chance also has an a priori framework.

To be sure, velocity and position cannot be measured simultaneously in the microscopic realm. The moment one tries to measure the position of an electron, the velocity cannot be known, and vice versa. But if the position is not fixed, it is impossible to determine the velocity, and without calculating the velocity the position cannot be determined. Herein lies the truth of the uncertainty principle. There is simply no other way to confirm the microscopic universe except by statistical probability.

This discovery of something approaching chance in the microscopic realms of the universe is a fact. Nevertheless, when we come to the point where the macroscopic universe is constituted, where chance has probability, limits, and determination, we come to the law of cause and effect.

The high degree of physical freedom in the microcosm approaches the degree of freedom seen in the life world. Naturally, the details are different. Many physicists, like Arthur Eddington and Arthur Compton, thought for a time that the two were largely identical in degree, but their views met with strong opposition from their scientific colleagues. Still, the life world contains internal conditions with a degree of freedom that does not hold in the atomic realm.

The purposiveness aroused in the life world by instinctual urges interferes with the degree of freedom, but there is nothing like this in the atomic world. There everything is ruled by objective conditions. However, in opposition to the seventeenth-century mechanistic view of the universe, in which everything is determined, Heisenberg's recursion of an indeterminate view into physics has to mark one of the great milestones of twentieth-century quantum physics.

It is further interesting to note how "uncertainty" has found an established place. It is one of the reasons that Sommerfeld, Planck, and Pauling came, by way of a priori probability, to believe in transcendent conditions in the structure of the cosmos that precede all phenomena.

Chance and finality: the origins of cosmic evil

Phenomena that appear spatially as chance, are seen to possess a temporal orderliness, and those things that appear temporally to be happenstance are actually seen to preserve order through statistical laws. In short, events thought to be matters of chance in the cosmos are not absolutely so and must be said to represent "slips in transition" that are necessarily produced in a mechanized world through the categories of law, change, selection, and especially growth.

Because this necessity forms a criterion for purposive productions, it contains essential conditions.

The occurrence of such "slips" might seem scandalous. Some will no doubt demand that all things operate with mechanical perfection. But since there is high-level differentiation and a comprehensive degree of freedom, the possibility that not all the elements in all areas will line up has to be factored in from the start.

If such complications have to be taken into account, so do latent principles in the universe for repairing and restoring slippages that are recognized as cosmic evils. Such a belief takes us into the realm of religion.

Seen in these terms, it is precisely because of these slips that the creation of a new world through versatile combinations becomes possible.

In a discontinuous world cut off from purpose, an increase of versatility is absolutely incapable of assembling a creative world. What we would have is rather something like a profound thinker engaged in an inductive process to "rework a botched job."

Similarly, when we look at the tendency to mutation among living organisms, it is almost as if we were watching an inventor at work. Without discarding the record of one-time variations, those things are preserved which have adapted to the environment.

A chance event reproduced identically time and again is not chance. Even as mutations in living organisms break Mendel's law, they preserve a new version of the law. From this we learn that chance in the universe—very far from what our common sense would expect—can even keep *chance* itself alive. In other words, we are forced to recognize that chance has the power to come up with novel ways to adapt to new environments.

I do not mean to imply that all the misfortune arising from automobile accidents, industrial mishaps, and natural disasters is a wellspring of happiness. These things are certainly chance events. In fact, many of these chance occurrences are caused by things like physical exhaustion, psychological problems, and inadequate facilities which, if improved, could reduce calamities to a minimum.

Chance takes place where there is purpose, which is why some think that it is chance that gives birth to purpose. I have argued, however, that the essence of purpose absolutely eliminates the possibility of being generated by chance alone.

Any number of elements and conditions have to be provided for purpose to take shape. For example, the right amounts of sodium (Na) and potassium (K) have to be present in the human body. As we have mentioned earlier, illness results if the total quantity of Na and K does not amount to 0.08 percent of total body weight. There is a finality in the human body to safeguard this fixed quantity.

The fear is that these proper amounts might be compromised by traveling through the desert, flying the skies, fasting, or overworking. Invasions of chance can take place. The most unreasonable course would be to follow Darwin's way of thinking or the idea of Hermann von Helmholtz that all chance is due to mechanical imperfections in the universe. Helmholtz made the case for the fundamental imperfection of the eye. It seems to me, however, that the very fact that physiological images undergo a mental corrective to invert them so that they appear properly is a purposive event that transcends the physical realm.

It is precisely by way of chance that finality works. Chance is a fact, a phenomenon appearing in a purposive world as a necessary part of the unavoidable *slippage*. But there are limits to fluctuation and a priori conditions given to chance. Further, correction is not impossible, but *slippage* and variability may even open up into the creation of novelty.

Mechanism and purpose

Mechanical purpose

In the second half of the nineteenth century, when Darwinism was in full bloom, the structure of organisms was held to be the result of purposeless mechanisms, changes in which were thought also to be a matter of chance. From where we stand today, such radical ideas seem to have been subscribed to rather casually, yet even among contemporary scientists there are those not given to philosophical reflection who have no trouble continuing in this line of thought.

Darwin himself ended up at a loss over the relationship between natural sifting and teleology. In his words, he felt like "an architect trying to build a house out of stone in its natural state without cutting it."[3] Hence he knew it was sheer nonsense to think about architects or building without allowing for blueprints and design.

Darwin was stimulated by Helmholtz's work on "imperfection in the eye," which for him meant that there is no design or purpose behind the lack of perfection. We are left reading him with a sense of desolation.

Even where design is present, there is no reason to expect absolute perfection. In a limited space, it is unreasonable to ask for anything more than a limited machine of the greatest possible size. More importantly, there is a prearrangement to be seen in the crude qualities and apparent imperfections of the human eye that supersedes its limited perfection.

The human eye that inverts what it sees in a single lens is physiologically imperfect. And yet mentally the imperfection is offset and there is an inherent versatility in the ability of the eye to make use of other kinds of lenses. If the construction were *too* perfect, the human eye that recognizes frequencies in spectrographic wave lengths when looking through a microscope would be useless in observatory telescopes, which require measuring of distance in order to distinguish differences of distance. Today it is the very simplicity of the human eye that has come to demonstrate its perfection.

The idea that imperfection argues against purpose is fundamentally flawed. Evolution, imperfect at the start, is little by little being perfected. There is not the slightest hindrance to psychological compensations for physiological imperfections or social morality filling in the gap for psy-

---

3. E. B. Poulton *Charles Darwin and the Theory of Natural Selection* (London: Cassell, 1896), 111–19.

chological imperfections. This is all made possible by the determination of purpose. It is precisely because of this purpose that our physical visual apparatus has been able to develop as we have it today. To doubt the connection of that apparatus to a purpose is to doubt the existence of the apparatus itself.

The disparity between human purpose and mechanical purpose

The difference between purpose in machines and human purpose is vast. Humans possess internalized purpose. They can set themselves goals as they wish, because purpose lies within. Machines, in contrast, have their purpose outside of themselves; they cannot determine their own purpose. Those who made the machine decide on its purpose. This does not mean that such extended purpose ceases to qualify as purpose. Because machines are machines, they can never lack purpose.

Why is this so? Many scientists claim that it is because the universe is mechanical but without purpose itself. This is my reason for connecting the purposive character of machines to a planned design. When we see steam engines or electrical equipment set up in a carpentry shop, we immediately understand that someone has planned for them to be there. But when we look up at the heavenly bodies as machinery set up in the world of nature, or at something as ingenious as our own human bodies, we dismiss them as products of nature that came about blindly and without a planner.

We admit to natural generation in continuous things, but then treat the atoms as all discontinuous. If we assume that the whole of the changing world moves blindly and by chance, then how do we account for the fact that atoms bond to carry out a unified function? Can this be chalked up simply to laws? If so, how are we to describe those laws? I call them purposive. A law is a closure of choice allowing for a single change. Choice means that there is something to choose between. The relationship between chooser and chosen is what I define as purpose.

Darwin acknowledged selection but denied purpose. I consider this a fundamental contradiction in his thought.

In any event, a machine can only come about as a union of two or more discontinuous parts. For it to operate, it must be infused with power dependent on a purposive selection *for the sake of* which it is operated.

The mechanisms of the natural world are all too grand and subtle. This is why so many of its parts are incomprehensible. Some, like Emil Heinrich du Bois-Reymond, end up in agnosticism, but there is not the

Cosmic Purpose

slightest need to deny purpose simply because of an abundance of grandeur and subtlety.

The destiny of the machine

None of us use our heads like machines to select materials and assemble parts. Machines do not permit chance. Machines are based on efficiency. Is there anything as efficient as the work of enzymes hidden within the particles of cell mitochondria present in humans and other living organisms? Have human beings ever invented any apparatus as excellent as the electrode activity of hydrogen ions (pH) in proteins? Within the universe, the mechanical operations of the natural world come about through the duplication of several hundreds or thousands of selective activities. If the claim is to be made that this activity is without purpose, the fact that it takes place must be recognized as something beyond chance.

Machinery set up in a factory represents a determination of laws that are: (1) reversible, (2) cyclical, and (3) not accompanied by development or progress. Purpose, however, admits of progress and development. This is one reason for rejecting purposiveness in machines.

On closer examination machines as such do not lack development altogether. A certain job has a sequence in which it must be done; this is its arranged orderliness. It moves ahead progressively. The assembly line is exemplary in this regard. Directionality is determined, position is taken into consideration, and a progressive development to the final stage is arranged through the application of power. And yet the fact remains, the development of autonomy found in life is absent in the machine because of the periodic cycling required for it to carry out its purpose. Nonetheless, it is not that living organisms are totally lacking in the kind of cyclical metamorphosis without advance or development, such as that seen in the lower animals.

Yet even here, we cannot presume to deny purpose.

In the vast cosmic machinery that seeks to conquer the future in order to complete its purpose and adjust to change, a continuity of energy linking the past to the future is required. This continuity of energy was one of the three great postulates advocated in the nineteenth century through a commitment to the principle of the conservation of energy.

This principle states that the cosmos is continuous and that energy constitutes an irreversible and repeatable cycle. Today the principle of the conservation of matter has in fact been overthrown, and the general claim that energy, which is equivalent to matter, is never depleted is

Knowledge of Cosmic Purpose

rather risky. Within the mechanisms of the earth's surface, the principle of the conservation of energy with its accompanying conditions arises together with the principle of the conservation of matter. Since that claim is only partial and does not cover the outer reaches of the universe, it is tantamount to a religious belief. In any case, we can be grateful that a continuity of energy has been arranged for things on earth.

If the completion of evolution is taken as the purpose of the universe, the continuity of energy requires it to be mechanically irreversible. Much like goods manufactured on a conveyor belt, the workflow is distributed with a destination in sight and is repeatable, but the conveyor belt itself is always irreversible.

Components with this kind of continuous mechanism are necessary to realize a goal.

We have already described the need for mechanisms if space and time are to be overcome and converted to a purpose, but insofar as mechanisms are involved, the opposite may also be said. In other words, given their mechanical nature, both space and time can contribute to the realization of a goal, but for this to happen, directionality and coordination have to be provided. Efficiency is possible only if mutual connections are established between directionality and coordination. Gravity, too, becomes a kind of motive power when applied to the mechanisms of a digital clock. Gravity is never present by chance, nor does it function blindly. Surprising laws lie hidden within it as a rationality that makes them by nature purposive.

The human body has its own mechanism. Its internal organs are all driven with directionality. Their development stops at a certain stage, progress is impeded, and everything thereafter runs mechanically on a single track. It is through such a mechanism that the body is able to realize personal goals.

Thus the mechanisms of the natural world play a transitional role in the realization of purpose. They appear with purpose and give it transit.

Mechanism and freedom

It is an error to think that because something mechanical involved is, all freedom is taken away.

The mistaken idea that machines have a fixed destiny and possess no freedom of choice when it comes to purpose stems from insufficient study of the degrees of freedom that appear in the mechanical structures found throughout the world of nature.

The natural world exhibits at least seven kinds of freedom: (1) freedom of movement, (2) freedom to change direction, (3) freedom to convert coordination, (4) varieties of freedom of choice associated with the physical chemistry of living organisms, (5) freedom of growth, (6) freedom to replace laws, and (7) the synthesizing freedom of the nervous system.

Freedom of movement. The degree of freedom of variation involved in melting points shows that even thermodynamically, the predestination of the natural world does not eliminate freedom altogether. In thermodynamics, the degree of freedom is expressed in the following formula:

$$F = C + 2 - P$$

F is the degree of freedom; C is the number of different constitutive chemical elements; P is the number of equilibrium surfaces in chemical equilibrium; and the integer 2 represents the factors of pressure and temperature. If we add to these electricity, attraction, and magnetism, the number is raised to 5. Simply put, this gives us:

change = things that change − things that do not change

Change occurs here in correlation to the degree of freedom. If the degree of freedom is 1 and the degree of indetermination is 1, an element becomes unstable. Besides the thermodynamic degree of freedom, the range of movement is correlated to factors like distance, velocity, oscillatory propagation, reverberation, and radiation. The range of movement is limited by these factors and yet maintains a certain degree of freedom.

Freedom to change direction. According to conditions of attraction and increase in velocity a circle is converted into an ellipse, the movement of the ellipse into a parabola, the parabola into a hyperbola. These conversions naturally entail different directions.

Colloidal chemistry demonstrates variations depending on differences in the amount of matter and density of electrical ionization.

It is a well known fact that if we look at electrical currents and magnetism, a conversion of direction results in a conversion of movement.

Freedom to convert coordination. If we acknowledge a degree of freedom present in the conversion of energy levels in paths of electrons within the atom, or in the reorganization of the spatial grids of crystalline bodies, we can also speak of a freedom present in the conversion of atoms into the three forms of solid, liquid, and gas, as well as in the variability of electrons to take the two forms of matter and light.

Or again, the shifting of the center of a crystal to the body, or the transformation of isometric crystalline forms into square or rhombus forms must also be said to represent a certain degree of freedom. We usually think of physical laws as all determined or prearranged, but it is interesting to see how freedom remains within the otherwise predestined realms of the natural world.

When we look at muscle movements biochemically, we see that the conversion of coordination through lactic acid causes an expansion that brings about movement within the muscle itself.[4]

Freedom of choice in physical chemistry. Regulations governing the choice of energy levels by electrons within the atom completely ignore the laws of chance, demonstrating a quality not unlike the thought processes of the human brain.[5] If we grant that the same kinds of selectivity present in electricity and magnetism are all due to atoms, the presence of a freedom of choice vital to the nature of chemical combination is immediately obvious when studying the chemical compounds.

Have we not seen the extent to which the freedom of variability shows up in the organic realm with the addition of hydrogen and hydrogen ions, or again, the extent to which change is present in the freedom of the organic world to adapt through the appearance of oxygen and water? We have to admit that the drive toward the potential emergence of life centered on carbon or nitrogen marks a great revolution within the physicochemical realm of mechanical predestination. All these things need to be seen as the momentous creation of a natural world traveling a straight line along the path of freedom of choice.

Freedom of growth. Whether colloidal magnesium is present or not makes an enormous difference to growth in plants. Similarly, the growth of animals is greatly affected by increases or decreases in the amount of vitamin B_2 or flavin. Thus the extent of the influence of physicochemical elements on growth has to be considered a major indicator.[6]

In addition, we are surprised to learn through recent research on plants and animal hormones that biochemical hormones determine the range of growth for all living organisms.[7]

---

4. See Hashida Kunihiko 橋田邦彦, 『生理学要綱』 [An outline of physiology] (Tokyo: Tomikura Shoten, 1943).

5. *Ibid.*

6. Linus Pauling and Samuel Goudsmit, *The Structure of Line Spectra* (New York: McGraw Hill, 1930), 77.

7. See Yamao Yasumasa 山尾泰正, 『ホルモンの探究』 [The study of hormones]

Cosmic Purpose

Freedom to replace laws. Even if we consider the realm of physical chemistry to be determined and predestined, this does not mean that it is completely bound to its laws. The fact that things that tie electrons to a circular orbit can supersede the utility of their limitations and be liberated to form light waves, has required a replacement of the laws of quantum mechanics, something that until now made no sense to physics.

Likewise, when we come to the interior of the atom's nucleus (10^{-12}), we find ourselves in a world where the laws of quantum mechanics no longer apply and have to be exchanged for new laws. But, unfortunately, humanity has not yet discovered all of the laws of nuclear physics.

Without entering into the question of the replacement of laws at the microscopic level, the need for substituting or overturning laws at the macroscopic level is clear from the development of the principles of relativity by G. F. FitzGerald, H. A. Lorentz, and Albert Einstein. Simply put, laws themselves may be absolute, but the world of matter with its variable character is in constant flux and subject to change. This means that along with fluctuations in position and changes in velocity, laws have also to be replaced.

Solids at a temperature of -273° lose their binding power. They become brittle and quickly fall apart. And what happens after that? The question has not been subject to enough experimentation, but there is no other path open for it than to convert to light. Thus, at -273° the laws of Gay-Lussac come to a dead end and have to be replaced with optical laws.

The same may be said of FitzGerald's law of contraction, according to which a physical object moving in a fixed direction contracts in that direction. If it is propelled at the speed of light, the direction in which it runs becomes zero, and only things perpendicular to that direction remain as width. In fact, propelled at the speed of light, atoms that constitute matter cause vibrations and disperse into "St. Elmo's fire." Meantime all sorts of transformations occur within the nucleus of the atom that would make an object burn up like a star. Hubble's research at the Mount Wilson Observatory reported on stars moving close to the speed of light at 120,000 miles per second (the speed of light being approximately 186,000 miles per second). The fact is, these are flat nebulae, but if they were not stars they could not be propelled at such speeds.

In any event, despite the fact that laws are determined, endless variations occur in the realm of matter and this means that the laws have to be

---

(Tokyo: Kōshi Shobō, 1943), 55–72.

replaced regularly to respond and readapt to changing conditions. This is why laws that are predestined retain the freedom to shift at times from one destiny to another. And just as a conveyor belt attached to a machine can be adjusted to move forwards or backwards, a completely new freedom is acquired through the exchange of laws.

The synthesizing freedom of the nervous system. Once the previous six freedoms—movement, direction, physicochemical selection, growth, and law replacement—have been acquired, it is not insurmountably difficult for the nervous system to bring about a general synthesis and convert them into psychological freedom.

Various forms of the degree of freedom have been arranged for even within the apparently predestined framework of an apparently mechanical world. Through all sorts of configurations of these degrees of freedom and within a framework allowing for mechanical organization, the nervous system seems to me an outstanding domain of free will. Far from doing away with the mechanical, physicochemical structure, it builds on its foundations to produce the wide berth of freedom in a continuous framework that we find in the cerebral nervous system.

Discontinuity and freedom

In spheres of discontinuity, change is strictly controlled. Hence, in a world where variables are considered a matter of chance continuity, the capacity for mechanical assemblage can only seem miraculous. The natural world works just such miracles. I have therefore stated in my treatment of chance and purpose that change and chance are never opposed to possibility; rather, they are a guide to it.

The categories governing the natural world do not stand independently and on their own. Change, growth, selection, law, energy, life—all of these are configured, fused, and woven together into something very similar to the activity of the human spirit. This analogous activity and an even greater ingenuity need to be recognized at work beneath nature and within it.

The assembly of mechanisms cannot be understood without a foundation in variation. At the same time, if we take only variation into consideration, certain forms of continuity become impossible to sustain. Already here, we see purpose come to birth.

The generation of something fixed within variation is not possible without purpose. Even so, many scientists take purposiveness to mean an absence of law, order, and structure, and therefore shy away from the

Cosmic Purpose

introduction of teleology into scientific research. However, I am painfully aware that it is even difficult to hold to a mechanical world-view with no mention of teleology.

Through some marvelous power of adjustment, dynamic mechanisms permeate all change and give continuity to movement. I have already explained the presence of purpose in this continuity. It may also be apprehended in the study of celestial bodies. Among variable stars, those that are not binary or variable stars are not able to give off light approaching that of fixed stars without going through adjustments in various sorts of atomic ions. When only the hydrogen ion is present, the light is unable to return to its source without a long-term cycle. Those that glow by means of elemental ions like beryllium and boron restore their light intensity in short periods of time. Fixed stars like the sun are not without cycles, but their adjustments are made by means of carbon and nitrogen ions.

The simple continuity of light intensity is made possible through a complicated assembly of atomic ions. When it comes to the great question of the permanent conservation of energy, things are still more complicated. The idea that there is no purpose at all in these cosmic mechanisms and that, however ingenious their movements, it is all a matter of determination, overlooks the fact that they are the result of a durability that penetrates these changes.[8]

As is the case with many stationary waves, matter entails a mechanism that concentrates several hundred and thousands of parts with dynamic quanta. These concentrations all come about with an a priori selectivity, as Sommerfeld and Pauling have pointed out.

Yes, a priori selection! Without it, there would not be a single atomic particle! One atom of uranium represents a regimentation of about 238 protons, neutrons, and mesons as well as 92 negatrons. It is absolutely unthinkable that a system with such a priori selectivity could be a product of blind chance. Obviously, such mechanization is more than predestined. It represents a realm with ample room for variation, choice, and all sorts of degrees of freedom.

The essence of cosmic purpose

Sustaining even one atom is far from a simple task. The world is too com-

---

8. See George Gamow, *The Birth and Death of the Sun: Stellar Evolution and Subatomic Energy* (New York: New American Library, 1952).

plicated and too ingenious to dispose of with old-fashioned, simplistic materialism or primitive mechanistic theories.

Without a certain physical degree of freedom, it is most difficult to maintain a prototype. Selectivity germinates in the attachment of limited conditions to the degree of freedom; within that selectivity a tunnel of cosmic purpose is opened up.

The actuality of purposive law

The existence of purpose was not acknowledged in the evolutionary theory that developed during the nineteenth century. The universe was held to be a blind, chance combination. I have my doubts about this. Is the world of sifting, sorting, and selection really that blind? Darwin in particular held to the durability of chance, but had no trouble seeing the domain of law beneath it. Even today there are some who simply accept that chance and law coexist without bothering to think about it. Things are not so simple in reality. What seems to be chance is subject to something like Heisenberg's principle of uncertainty. Chance has its limits, on the other side of which lies a domain where things have been decided.

Even from a purely epistemological standpoint, the idea of the universe as a continuity of chance is preposterous. If chance were responsible for the birth of law and selection, then it would have to contain these things within itself. That cannot be. I acknowledge that change occurring apparently by chance exists in the universe. Without *change* probably nothing at all would exist. Energy is born of *change*. There is no mass without energy and no matter without mass. We have not the slightest reason to doubt that matter was born of *change*. But change comes with conditions and prearrangements attached. There is a *degree of freedom* in energy which is bound to laws. Water has three degrees of freedom: ice, water, and steam. It would seem to be chance that brings these degrees of freedom together.

To Darwin's way of thinking, things that exist in the natural world are based on law and sifting. He considered change, selection, and law the conditions of existence, but was unreasonable in his neglect of purpose. Later he noticed that the "local purpose" of what he called the choice of mates is built into the world of life. He might have insisted that the choice of mates is also the result of chance, but once he had differentiated the life world into male and female, he could no longer claim the intervention of chance.

It is not that there are no philosophers who hold that values are completely distinct from existence, as if existence were there before values showed up. But I agree with the existentialist Martin Heidegger that consciousness of existence would be totally impossible apart from the existence of the knowing human subject itself. In our human *Existenz*, being and value are completely one. The same may be said of objective existence. Darwin based the existence of life on the three great categories of sifting, law, and change, but these categories are a function of human consciousness. We must therefore acknowledge the dominion of laws common to the world of matter and the world of mind. Many materialists accept the three basic principles of change, selection, and law but deny a principle of purpose running through the material world. But I find myself having to recognize that along with the other minor categories, a principle of purposive law belongs to the law of existence. There are five reasons for this.

1. Knowledge of existence is not possible without taking into consideration a purposiveness that assumes knowledge of an objective purpose for knowing existence. I call this "logical purposiveness."
2. Objective existence between the nothingness of the past and the nothingness of the future is given an existence known as the present. I hold that this existence comes about through a particular selectivity. Given that the present, bound on both sides as it is, is not a chance existence, we may consider it a purposive existence made manifest through selectivity.
3. Without taking cosmic purpose into account, the distribution, integration, and management of energy with its tendency towards the evolution of life and material bodies would not in fact be possible. This is because distribution, integration, and management imply selection. This discussion grows out of the recognition of purpose in life.
4. I recognise that the presence of phenomena like unity, harmony, design, order, adaptation, beauty, and sifting in the natural world presuppose the existence of purpose in nature. This is the argument for purposiveness in nature.
5. In addition to the above, from the emergence of the personhood and consciousness of the knowing human subject, I hold that purpose is an actual fundamental principle rooted in the universe. This affirms the actuality of conscious purpose.

The existence of purpose from the viewpoint of epistemology

Truth arises from judgment. Sorting out the truth may be relative in the sense that there would be no truth if there were not also falsehood. But truth is not a result of chance. A book is lying on my desk. The fact that it is made up of many pages constitutes an absolute truth within a relative world. Perhaps someone placed it there. But if the book is here on my desk, my intuition needs no supporting proof to confirm this fact. There is no denying purpose is present in the eye looking at the targeted object, the book. It is not that an "eye" that has come about by chance is looking at a book that has come about by chance. The reflection of the book in the eye did not just "happen"; it is the action of a human being for the purpose of knowledge. Furthermore, no doubt the book was purposely written by an author. Naturally, its printing had to enlist the help of a machine, which also contributed to the existence of the book.

Hence, in the sphere of knowledge, we cannot even begin to think of existence in isolation from purpose. Purpose also exists just as assuredly in the universe itself.

Purposiveness in existence

The astrology of ancient Babylon and Greece held that each appearance of the planets and comets had a kind of meaning and purpose for humans on earth. The teleology I have in mind is of another sort. The movements of the stars are mechanical. In the end, there is no need to deny that these mechanical operations control the environment of living things, that directly and indirectly they cause air pressure, ionization in the atmosphere, earth magnetism, earthquakes, and so forth; nor that evolution in all its shapes and forms on earth depends on those things. What I cannot accept is the realization of purpose through machinery.

I consider it a mistake to hold that the whole world is connected only by way of direct purpose. In other words, it is wrong to reject *progressive* purpose in favor of some *personal* purpose behind all things. I see at least seven types of purposiveness:

1. Purpose related to energy shows up already in the phenomenal world with directional purpose and its more complex form of coordinating purpose.
2. By means of adaptational purpose and its more complex form of structural purpose, change brings about the systematic organization of organic bodies and crystalline bodies in the atomic realm.
3. In the area of growth, progressive purpose and its more complex form of catalytic purpose are active in the phenomenal world.
4. In the area of sifting, limited selective purpose and its more complex form of discriminating purpose appear.

5. The realm of law sets up decisional purpose, which closes in on a single choice or a single decision involving selection, and its more complex form of the contractual purpose involved in cause-and-effect relationships having to do with the distant future.
6. In the case of the world of life where unity and harmony among these five forms of purpose is preserved, impulsive purpose appears as desire. An infant is given an amazing instinct to suck at its mother's breast without having to be taught, thus fulfilling a wondrous purpose.
7. Despite the presence of conditions related to causes and effects that borrow from other energies in the same system, as the above six types of purpose pass through life and aim at becoming conscious, self-determining purpose is given structure. As conscious purpose with the flexibility to adjust, it approaches the actuality of the cosmos. The actual existence of matter is made known to this consciousness.

This sevenfold division of forms of purpose is extremely significant. Conflating the seven makes it impossible to understand the true meaning of purposiveness.[9]

By conflating directional coordination with conscious purpose, astrologers generate the most dreadful kind of superstition with truly distressing results.

In the case of the movement of energy, can we understand kinematics without taking motion, angle, and velocity into account?

Geometrical electron optics has succeeded in creating an electron microscope without a lens. The rationale is exceedingly simple. A lens ultimately makes use of the direction of light rays in carrying out its various functions, but eliminating the lens that actually determines the direction of electrons and making use of electromagnetic energy instead allows for a more powerful electron microscope.

A certain purpose is achieved simply by using direction and nothing else. Those who do not stop to think about this very deeply are not likely to appreciate the grandeur of this purpose. Nevertheless, directionality on its own has a great role to play in the production of an electron microscope.

Similarly, there is already latent purposiveness to be found even within the realm of energy with its apparently complete absence of purpose. Would the amazing kind of autonomous self-determination of conscious

9. See page 193 above.

purpose we see in human consciousness have shown up in the universe without it?

Even more to the point, as Max Planck has taught us, energy is based on the minute energy of quantum-photons smaller than the electrons of "Planck's constant" (655×10^{-27}). Integral multiples of these quanta forming neutrons, mesons, and protons make up the ninety-two chemical elements present on the earth's surface, which are built up only of integral multiples of the elements, protons, and so forth that constitute them. The appearance of integers in the atomic world does not allow us to subscribe to the rule of chance as nineteenth-century scientists had.

This simple explanation is not, of course, sufficient. An understanding of the natural phenomena recorded in these pages obliges us to recognize that existence is not simple existence but is based on an actual purposive principle.

Absoluteness in Purpose

We need a new standpoint from which to view the absolute nature of purposiveness.

1. Purpose synthesizes duality into a unity, as in $A't' \to A^2t^2$ where the space-time duality is unified.
2. Energy, change, growth, selection, and law are all synthesized in purpose.
3. Subject and object are synthesized into a unity in purpose.
4. Purpose synthesizes the objective energy of subjective selection.
5. In positing purpose as a shift from *nothingness* to *being*, the true absolute nature of the creation of being from nothingness is realized. Dialectical variation in the being of objective being and subjective being does not produce a great being, but the absolute is actualized as being within nothingness.
6. A selectivity obtains that could not be actualized if the being of purpose were not absolutely prearranged in the realm of nothingness.

Monistic plurality

A monistic plurality can only be explained through purpose. Of necessity, purpose begins in objectivity, is motivated through subjectivity, and through the pursuit of practice effects a synthesis of subject and object, or in other words, advances to a world imbued with purpose. These boundaries mark the realm of an "absolute."

The content of purposiveness

In addition to a wooden lever's use in prying an object loose, there is also purpose in the simple stick of wood cast aside in the shed. A heap of old rags tossed into a drawer each has its own purpose *attached* to its existence.

The term *purpose* immediately suggests human free will or self-determination, but purpose within the cosmos cannot be reduced purely to self-determining purpose. Purpose has a mysterious composition that, like a jack-in-the-box, can pop up in any number of ways. Each and every appearance has its own kind of purpose and strives to realize a grand purpose.

With extremely simple things, we immediately recognize what kind of purpose is involved, but when it comes to more complex things the purpose is often hidden to us. For example, we understand at once that the purpose of the small intestine is to absorb nutrients, but the purpose of the appendix has not been identified. In lower animals the appendix has an important task to perform. In some cases it serves as a sack to hold protozoan bugs. The human appendix may have a mediating role to play in development from the fetal stage, but to this day that role remains completely unclear to us.

TYPES OF PURPOSE

Progression in purpose

Purpose is necessarily accompanied by movement. Spatially speaking, purpose has to do with linking two or more positions, and temporally with linking two periods of time, but neither can it be realized without movement. It is difficult to discover the purpose of an entity from one part cut off from the rest or from only a simple, static body.

The ribbon in which a gift is wrapped has no purpose on its own. A recipient would be angered to be handed only the ribbon. Still there is a difference in purpose depending on whether the gift is wrapped in a ribbon or not. The ribbon is a symbol of the courtesy shown by one party to another. The same is true of a flag, a drill, a saw, a desk, or a brazier. None of these has a purpose by itself. Its purpose exists in its progressive movement or in the meaning it helps to convey. Without progression it is difficult to discern purpose. Accordingly, the types of purpose can be distinguished according to the range and duration of this progression.

Seven types of purpose

To reiterate, I divide purpose into seven sorts: directional purpose in energy, adaptational purpose in change, progressive purpose in growth, selective or sorting purpose (accommodation), contractual purpose in law, impulsive purpose in life, and self-determination in the realm of purpose.

Each of these types of purpose is combined with, connected to, and developed with a complex of other purposes. Directional purpose shows design in its connection to a multiplicity of other directional purposes, coordinations, inclinations, and so forth. Adaptational purpose is tied to many elements involved in adaptation, suggesting an organizational purpose. Progressive purpose shows catalytic purpose in preserving connections to other things. Sorting purpose discriminates among many elements to become discriminating purpose. Through the combined arrangement of the individual contractual purposes of cause and effect we have things like hereditary factors in living organisms.

The life world manifests genetic and organizational purposes that anticipate a more unified arrangement.

We may summarize purpose in these seven realms with a diagram:

1. Energy ………… directional purpose ………… design purpose
2. Change ………… adaptational purpose ………… organizational purpose
3. Growth ………… progressive purpose ………… catalytic purpose
4. Selection ……… sorting purpose ………… discriminating purpose
5. Law …………… contractual purpose ………… genetic purpose
6. Life …………… impulsive purpose ………… instinctual purpose
7. Purpose ……… self-determining purpose …… self-conscious purpose

DIRECTIONALITY AND PURPOSIVE DESIGN

Energy and purpose

Whatever the differences in space and time, there is no movement without a determination of direction. What actually determines direction differs from case to case, but the purpose of energy is also a response to its location and is known through its direction. Alternating electrical current has two directions. When there is a need to adjust to direct current, selenium, germanium, silicon, mercury, copper oxide, and the like are used to make it unidirectional. Here a single purpose is given in its direction. The pulsating current that occurs within plant stems clearly has one purpose, just as the osmotic interchange in intestinal mucosa shows a clear latent purpose.

Thus clear direction of movement accords with a teleology. The organic chemist Wolfgang Ostwald pointed out that matter as a whole has a single direction, which, while not patently a *purpose*, can be interpreted as steadily advancing in the direction of value. Through his research on energy levels in atoms, Sommerfeld further points out a selective purpose in each atom.

Nevertheless, I do not base my views solely on such specific research. Rather I should like, from a more general standpoint, to inquire into the design of energy in the universe.

Design in the determination of a "field"

The most astonishing thing in physical chemistry is the determination of a field in the realm of energy. The microscopic realm is comprised of the mystery of the quantum discovered by Planck. According to his research in thermodynamics, heat radiation does not maintain continuity through oscillation. We cannot speak of radiation before reaching a fixed group (h), but radiational activity first occurs at a certain integral multiple of h. In physics, this h is known as "Planck's constant."

The discontinuous "field" of energy activity, which only occurs in integral multiples of a constant, leads me to think that there is not only direction in energy but also a coordinating design that runs through the field. We observe this taking place when there is an increase in directionality.

These quanta represent the world's smallest energy fields, but when enlarged to the realm of the electron, they develop into the atomic realm.

As we explained earlier, Henri Moseley took an important step in the field of spectrochemistry with his discovery that atoms are arranged by frequencies corresponding to a formula of the number of light waves.[10] The fact that this allows for all atoms to be included in a fixed formula represents an amazing, monistic world of order.

Molecular fields and design

I have spoken briefly of design in the atom, but there is also an amazing design of latent purposiveness in molecules, as the pioneer of electromagnetic theory James Clerk Maxwell noted. If we examine the

---

10. See page 72 above. We might add that the new quantum mechanics thinks of electron frequencies as "clouds," but this does not pose an obstacle to older approaches.

"field" constituted by the three-dimensional coordination of organic compounds, beginning with water, oil, and proteins, we immediately understand the latent purpose running through the field. Molecular formulas may differ, but their influence on human bodily function is absolute. Humanity has produced all sorts of medicines by studying the arrangement of molecules.

The most surprising of these is the structure of water. Study of the axis of the simple chemical compound H_2O makes us ponder the miracle of its molecular field. How do flammable hydrogen and oxygen become non-flammable? Why should water, as a carrier of electrical charge and as something with a stunning capacity for expansion, show such an extreme degree of solubility?

The field of celestial bodies and design

The activity of the *fields* associated with quanta, light, electricity, antiprotons, mesons, neutrons, positrons, neutrinos, and other minute particles is such that, outside the limits of the *field* no activity is possible. Yet when several of them are brought together and organized, they form a new field, becoming atoms, molecules, cells, living organisms, displaying self-determination in a world of finality. These miniscule phenomena on the earth's surface can also be applied to the macrocosm of astronomy.

The distribution of the planets in our solar system is arranged in integers. This is known as "Bode's law," after its discoverer.[11]

Hence research shows that there is a relation between the direction and field of energy, and that the movement of energy within the universe is never meaningless. The movement of electricity in an electromagnetic field is perpendicular to the field opened up, and the magnetic field for the electric current has the determining quality of an a priori probability.

The design of a field

The amplitude of a vibration is also seen at once to have the determination of a field. A quantum has a field of 10^{-27}, an electron of 10^{-13}, and an atom of 10^{-8}.

These realms of oscillation are obviously of such a nature that they will have to disappear. Nevertheless, the fact that the structure of the earth has a form approximating the total conservation of matter gives us pause

---

11. See diagram on page 80 above.

to consider how well designed it is. As "wave," matter never disappears. Thus the energy that shapes matter should never disappear either. But when we observe matter within the "field" we call earth, it has an amazing "immortal" kind of structure. We have to stare with awe at the engineered design that leads to a form of matter that continues on forever, packed within a fixed field of wave motion. As Henry Norris Russell has pointed out,[12] this undying form of matter makes us posit a relationship between the mass of the earth and its diameter. If hydrogen can never escape from earth, neither can other fundamental elements. Because of this wondrous structure, all matter on the earth's surface has a certain immortality, and energy, relatively speaking, also has an immortal quality. This guidance from mortality to immortality, together with the rational arrangement in integrals of continuous oscillation into discontinuous particles, obliges us to create a new, teleological view of the cosmos.

In his book *Man, the Unknown*, the celebrated French physicist Alexis Carrel reported on research into the finality of blood. He was astonished to find that with each division, white corpuscles, which were thought to have only a simple structural activity, display the capacity to generate new blood vessels within which the blood can circulate. This, he says, makes us think that even physiologically speaking, there is finality in the universe.[13] Here we detect a new direction for science.

The many faces of purpose

Adaptational purpose and organizational purpose

The simplest things, from a single object to a single movement, possess adaptability. This is how I define adaptational purpose.

When many parts with adaptational purpose are put together, a complex machine results. I call this organizational purpose.

For example, no one of the bone fragments or ossicles that make up the skeletal structure of the hand is able by itself to make the hand move smoothly. But when the 12 bones of the palm are arranged in 3 rows of 3, 4, and 5 respectively, and then lined up above the 2 ulnae and radii, which in turn are placed above the humerus, they are ramified in an order—1, 2, 3, 4, 5— that makes smooth movement possible. But if the ossicles that

---

12. *The Solar System and its Origin* (New York: Macmillan, 1935).
13. *Man, The Unknown* (New York: Harper and Brothers, 1935).

make up the hand were scattered, there would be no way of knowing, without specialized skill, which bone plays which role. Only when they are organized can their purpose be understood. I myself once labored in an osteological laboratory to put together the skeleton of a horse.

Progressive purpose and catalytic purpose

Adaptational purpose and its compounded form, organizational purpose, appear in change. When we come to time, we find an interim purpose that takes on a kind of intermediate role. This includes a progressive purpose that extends continuity to others. When progressive purpose combines with organizational purpose, we have catalytic purpose.

The Russian geochemist Verdnasky observed that iridium (iodine, atomic number 53) is distributed in extremely small amounts, leading him to think that it carried out a kind of catalytic action. It is truly interesting to note that in order to bond hydrogen and carbon in the manufacture of synthetic oil, elements such as cobalt, nickel, and molybdenum need to be employed as catalysts. With the discovery of the surprising catalytic action of rare earth elements, we have come to understand the hitherto unknown and amazing catalytic purpose that each of the atoms fulfills on earth.

Much the same can be seen in the interesting digestive and catalytic activity of microbial enzyme bacteria inhabiting the human intestines. Cheese can be manufactured through the catalytic action of the rennet bacteria that thrives in horse intestines. The Russian favorite, kumis, is made with microorganisms present in horse urine. Miso, soy sauce, and the niter soil under greenery all depend on a catalysis of microorganisms that human beings apply for their own ends. To look only at the purpose realized through material mediums used in alloys and dyes is to recognize the existence of purpose around the universe itself, and at the same time to see the presence of things whose purpose is to serve as a medium.

Of all the forms of purpose, the most difficult to discover are in fact progressive and catalytic purpose. Machines and apparatus are made with the quality of progressive purpose, but if we enter a workshop and are shown the array of manufacturing equipment, without knowledge of engineering we are unable to understand what purpose they all serve. The great workplace we know as the universe actually has the same sort of structure. It is just too large. The grand machinery of the solar system and the galaxies that belong to it are also too enormous and ingenious for us to discern the progressive purpose they hold.

The same holds for the inner structure of the atom. The delicate quantum and wave mechanical structure has a certain degree of freedom and selectivity in its make-up, but even with what we have learned from recent atomic physics, there are too few like Sommerfeld who has shown us the wondrous purpose in each and every atom. For my part, deep study of the mechanical structure of the atom has left no doubt that it possesses a kind of intermediate purpose.

Selective purpose and discriminating purpose

To convert alternating electrical current into direct current, the plus or minus is suppressed, allowing only one of them passage. This is the idea behind in the construction of an electrical rectifier. Copper oxide may also play that role, with silicon, selenium, and germanium carrying out the task.

It seems that internal organs with selective purpose are essential to the human body, as many lining membranes carry out osmotic activity fitted with selective functions.

Distinguishing microorganisms is a difficult part of bacteriology. Various apparatus and techniques have been tried out to that end.

Darwin, as we have said, claimed that "sifting" is an unconscious choice. This selective activity is represented by the survival of the fittest in which the weak are defeated by the strong.

Selectivity necessarily and by its nature attends the realization of purpose. It is involved in both adaptational and catalytic purpose. But inasmuch as physical objects do not include adaptability and catalysis as such, selective purpose means that the objects themselves bear the duty of sifting other elements out.

It is rather like a judge in a courtroom deciding on the merits of a case. Prison cells have the adaptational purpose of detaining prisoners; the jailors have the intervening, intermediary purpose of conducting prisoners to the courts; the judge steps in with a selective purpose. Or again, the locks and keys of the cell are manufactured on the basis of an adaptational purpose; the summons written up for a prisoner has an intermediate purpose; the prisoner's number tag is made with a selective purpose.

Contractual purpose and genetic purpose

Law indicates order in things like temporal development and spatial design. In particular, it indicates the contractual arrangements hidden within space and time.

Temporally speaking, a contractual arrangement has a single purpose. When one person says "Promise me!" to another, their contract points to the realization of an arranged purpose. The same may be said of laws in the universe.

Law means (1) the determination of a field for the movement of energy, (2) restrictions on change, (3) the guarantee of repetition in growth, and (4) a closure or determination of selection regarding choices to be made. As the closure of selection, law represents the opposite of the rule of chance. Temporally, law means persistence. Thus law indicates a complete order through energy, change, growth, and selection.

In terms of change, the discovery of lawfulness in space points therefore to a single design. Analyzed as a process of sifting, it points to restrictions on the degree of freedom. The realization of a realm of purpose, whether it be organic or inorganic, follows these determinations.

Countering nineteenth-century ideas of thermal radiation as continuous, Planck discovered that thermal radiation does not occur without a fixed quantity and an arrangement of integral proportions known as quanta. This determines a field within the realm of energy and at the same time indicates that there cannot be an unrestricted rule of chance in the cosmos.

Ostwald's "philosophy of values" has demonstrated direction in matter. From a spectrochemical viewpoint, Sommerfeld has shown us a kind of selectivity in the atom in a degree of freedom latent in the functions of the atom itself, and through that degree of freedom a fixed purpose in the lawful nature of the atom.[14]

We find the same thing in crystallography. Matter never exists in any form by chance. It is ruled by the law of "six forms and thirty-two states." Friedrich Fröbel, founder of the kindergarten, is said to have shown young children a crystal in order to teach them that it was no match for the design of the universe. Quite so.

If we look at the three-dimensional shapes of compounds in the organic world, there is no denying that proteins and other chemical compounds each has its own distinctive design. Confirmation of this can be seen in the recent discovery by scientists in Japan of over thirty forms of crystalline enzymes.

14. See note 31, page 102.

The laws determining the genes of living organisms, the basic factors of transmission, also give us to understand that law includes purpose.

Speaking of genes, the male-female factors within the cell number in the tens of thousands, but they will never form a living organism unless their minute molecules are linked one by one with a finality. This is a truly beautiful verification of how genetic law is joined to a complete finality.

The law of cause and effect, according to which a cause gives birth to an effect, represents one temporal contract. This arrangement harbors a "purpose," so that when a seed is planted a flower will always bloom or, after a period, bear fruit. The time has come for us to wake up to the fact that even though the purpose concealed in all laws may not be clearly grasped, as with the genes and their factors, through adaptational, progressive, and selective purpose along with each of their variants, the universe is filled with a consistent cosmic purpose in the form of cosmic laws.

Impulsive purpose and instinctive purpose

Nothing is as mysterious as cellular mitogenesis. A rift opens vertically in the center of the cell, which then divides it into four, eight, and sixteen parts, always in integral multiples of two. It then collapses inwards as the outer cortex moves to the inside to make a digestive system. A single cellular layer further breaks up into three, forming the inner, middle, and outer cortex. Various organs appear with the differentiation of skeletal, muscular, nervous, urinary, and circulatory systems.

Within this differentiation there is a unity through which development occurs and growth is arranged. Anatomists may view all of this mechanically, but there is a great difference between the impulse to life and mechanical function.

If a worm is cut in half, both halves go on living. If surgery is performed on someone to remove part of the stomach, after some years the stomach returns to its former shape. If you saw a bicycle or a locomotive in two, they will not move unless supplemented by parts from outside. Even if half the human brain is damaged, no great change occurs in nervous activity. In all of this we see the extraordinary difference between a mechanical structure and the world of life impulses and psychological instincts.

The question of whether or not there is finality in the life world has been a point of debate for many years, but with the arrival of atomic physicists like Sommerfeld, who discovered finality within the very mechanical structure of the atom, it was only natural that the shadow of mechanistic life theories would diminish. The point is clearer still if one looks at Hans Spemann's work on "organizers."

Earlier in this chapter I took up the structure of purpose, where I explained that without a mechanical constitution, the realization of a grand purpose would be absolutely impossible. Even when we come to the apparently exact opposite domain of free will, attempts to heighten the degree of freedom are futile unless the degree of mechanization is also raised. I subsequently noted the design in machinery and the presence of purposiveness. Plans have to be made before machines be built. If anyone thinks that the mechanisms that have appeared in the universe and been preserved in living phenomena for tens of billions of years are the result of chance, it can only be because of a deliberate refusal to face up to the incoherence of their logic.

Life impulses disclose a high level of purpose and, unlike machinery, need to take adaptation to the environment into consideration.

For a machine, the property of motion is decisive; generative functions do not come into the picture at all. For living organisms, beginning with the egg, there is generation, growth, and progress. When illness sets in, an alliance of cells steps up to repair and restore, without any outside help. This kind of finality is what Driesch called "entelechy." The idea is not as unreasonable as its critics assume. The discovery of *genetic factors* and *organizers* has partially reaffirmed Driesch's theory of entelechy in the strict sense of the term. Gustaf Strömberg of the Mount Wilson Observatory has argued that entelechy holds for the whole of the natural world as a capacity in genetic factors and systemic elements.

I cannot, however, subscribe to the claim of entelechy theory that the structure of finality can be reduced to adaptability. I rather think it should be seen as unifying directionality, progression, catalysis, discrimination, prearrangement, and so forth. Consider the division of a cell. A discriminating purpose differentiates inner, middle, and outer cortex according to directionality. Various organizational systems then interact to perform progressive and catalytic functions, each with its own purpose. Only thus, I believe, is life given its inherent and orderly functions of generation, preservation, and repair.

These functions are of course formulated in mechanistic, physical, and chemical terms. But scientists have assumed until now that if something is mechanical, it lacks finality. For me, a thing demonstrates finality precisely *because* it is mechanical. This is how I read Oparin's "material teleology."

Machines do not have self-determining purpose, but this is not so for the design purpose of those who lay out the plans for them. Machines have motion but not life. The living organism comes into being by the addition of the phenomenon of life to what is otherwise simply mechanical. From the viewpoint of the living thing, we may say that mechanical phenomena belong to the plans for life. Life is the master and machinery the servant. The master possesses a self-determination capable of repairing itself; the machine does not.

Finality and the evolution of the sexes

The evolution of *sexuality* is the most amazing phenomenon in the impulse to life. Spatially as well as temporally, males and females develop and reach maturity in complete detachment from one another. By instinct they discover a mutual finality and by instinct advance to bond into the purposive union of sperm and egg, culminating in the purposive union of male and female genetic factors. These three stages of purposive union differ from mechanical operations; this union belongs to a realm endowed with a degree of freedom that staggers the imagination. Even Schopenhauer, who put such great stress on the unconscious dimension, had to acknowledge in *The World as Will and Representation* that sexuality represented a kind of design.

The finality of the impulse to life is thus clear in sexuality. Indeed, all life impulses, aside from single-celled organisms, pass through this world of sexual impulse, which is why we can speak of the mechanical operations of life as disposed towards this gateway of finality.

Finality and the evolution of the placenta

When we study the evolution of the placenta in mammals, we see that the higher the animals, the more extensive the protection provided the fetus. There is no other way to explain this than teleologically. Kangaroos do not have a placenta. Like Australia's platypuses, they lay eggs, hatch them within their bodies, and suckle their young in a pouch attached to the outside of their bodies. In pigs, several fetuses grow tightly attached to a spherical placenta, but there is no external protective cortex. Bears and lions have cuff-shaped placentas like bottomless sacks, and the fetuses grow with their heads sticking in both ends of the sack. In cows, the placenta is like a skin that completely envelops the fetus. In the case of humans, the enveloping amnion has an inner and an outer layer that are filled with amniotic fluid, providing the fetus with complete protection. Contrary to what we see in wild animals, the extraordinary attention that is given the placenta in nonresistant animals as an aid to protection and growth is unthinkable without finality.

Finality and the mystery of the mammary gland

Finality in life impulses is also to be found in all sorts of other functions. The evolution of mammary glands and variations in their chemical substance offer the best proof of the finality of life impulses. Mammary glands develop in perfect proportion to the growth of the fetus. Milk is released from the nipples at the time of birth. Its nutritional value surpasses anything the most brilliant dietician could devise. From this we see that the finality of life impulses exceeds human intelligence. The purposive responses by which mother's milk increases in density as the infant grows, and by which mammary glands shrink when the infant is weaned, are entirely beyond the reach of our intellects.

The finality seen in the digestive system, circulatory system, nervous system, urinary system, skeletal system, and muscular system are too complex to touch on here, but the attempt to explain their functions in purely mechanical terms without taking into account the finality at work in the background cannot offer a suitable explanation of what is taking place in the world of nature.

The finality of instinct

The work of life impulses extends beyond the mechanical. I have shown that they involve the combination of all sorts of purposive activity, but we need also to note a finality in psychological instinct that runs parallel to biological function.

For a long time, many people have gotten used to thinking that "instinct is blind." In fact, it is quite the contrary. The tongue of a nursing baby is curved when it sucks at the nipple, creating a vacuum between the tongue and the upper jaw and allowing milk to fly out of the mammary gland and into the vacuum tube. Through what wisdom did such an instinctual devise come to the baby? The adult tongue is stiff and can no longer create such a vacuum between the tongue and the upper jaw, making it impossible to suck milk. Can such an instinctive finality be called blind?

Is instinct blind?

The idea that instinct is blind is often applied to romantic love. But there are enormous differences that make it a poor example of the operation of selection in the sphere of finality. It is precisely because instinct has lost its finality and become one-directional that people refer to romantic life as "blind."

Statistically speaking, the numbers of male and females born to humans and rodents in their ordinary environments is half and half, which speeds up population increase. With different proportions, the percentage of births would diminish. Thus, under normal circumstances the idea that sexual instinct is blind turns out to be unsustainable. In times of social, economic, and cultural unrest, serious deviations result in an uneven distribution of males and females, posing obstacles to selection and exposing the so-called "blindness of instinct." Instinct cannot be said, therefore, to be *fundamentally* blind. Just as we cannot deny a purpose to the spinning wheel because the spindle happened to fly off it, serious deviation can also upset the spinning wheel of instinct. This happens as a result of affirming a realm of purpose, without which there would be no need even to use the word *blind*.

Teeth serve the purpose of chewing food. When there is no food, this does not mean that the instincts of the teeth are no longer present. The same holds for sexual instincts.

The finality of motherly instincts

Finality is still clearer in the case of motherly instincts. The primary instinct of the mother is to suckle. The swelling of the breasts is in inverse proportion to the emptiness of the infant's stomach, so that when the infant requires milk the mother is ready to provide it. If she does not, the mammary glands can become inflamed.

In the case of white ants, the queen mother is confined for several years to a hold in the anthill to nourish the baby ants. The severity of the instinct is altogether surprising.

Consider motherly love in birds. When baby chicks hatch, the mother's love turns forceful, almost like someone ill-bred. When the mother cannot manage on its own, the male bird assists in rearing the baby.

Reading books like Henry Drummond's *The Ascent of Man*, and the works of Friedrich Alverdes and William Wheeler on social life among animals and insects,[15] we are surprised to find so much written on the motherly love of animals.

Alfred Russel Wallace has made interesting observations on the nest-making instincts of the Chinese francolin. Many of these birds instinctively build nests out of fish bones. When there are no fish bones to be had, they are not particular what materials they use for their nests. Hence their instincts are neither deterministic nor blind but exhibit the capacity to adapt according to the environment.

Jean-Henri Casimir Fabre was struck by the rearing instincts of the hunting wasp. Those in the philanthus group sort out ganglia in the larvae of their victims one by one and sting each of them with an anaesthetic poison—something not even the best comparative anatomists can do. The anesthetized insect larvae are stored in a hold as living corpses and when the eggs of the philanthus that have been spawned in their flesh hatch, they grow by eating the undecayed flesh.

Such instincts possess a finality exceeding human intelligence and display a surprising adaptability in an apparently blind world. In consideration of these facts, we can only extol the ingenious design of nature.

Self-determining purpose and self-conscious purpose

In studying the physical foundations of psychology, Jacques Loeb of California University noted that the behavior of lower animals shows a plantlike phototropism, negative phototropism, magnetotropism, and negative magnetotropism.

There is no denying that animals, by virtue of being organisms, are restricted by various biochemical conditions. On the basis of these restricting conditions alone, it would appear there is no self-determination at all in living organisms. There are even times when the movement of animals may be seen as a certain physiochemical response to the severest of conditions.

Because the physiological system of animals is made up of colloidal connections, it is flexible in the heat of summer, while in the cold of win-

---

15. Henry Drummond, *The Ascent of Man* (London: Hodder and Stoughton, 1894); Friedrich Alverdes, *Social Life in the Animal World* (London: K. Paul, Trench, Trubner, 1927); William Wheeler, *The Social Insects: Their Origin and Evolution* (London: K. Paul, Trench, Trubner, 1928).

ter it tends to chap, for which it is only reasonable that measures be taken in the form of a psychological response. Therefore all living organisms are mechanisms, but it is unreasonable to conclude that that is *all* they are.

We have already argued for an impulsive purpose to life and will not repeat ourselves here. The living organism, unlike the machine, has an enlarged capacity for self-alteration. When a machine is damaged it is not able to repair itself; living things can mend themselves and have the capacity for restoring themselves to their original form.

Living organisms, then, have characteristic powers that transcend those of machinery, but given the fact of phototropism and negative phototropism, we cannot dismiss out of hand the workings of reflex.

Such facts explain the truth of finality and cannot be turned against it. Nevertheless, this finality is impulsive.

Is it always the case with animals that free, purposive selection through self-determination is absent? Ants and bees build social structures for populations numbering in the hundreds of thousands. It is also a fact that they manage agriculture and manufacturing through a division of labor. Can we really explain all the activities of insect societies in biochemical terms? The answer is—no, impossible.

Determination of a field and variation within the field

Besides physicochemical reactions, such insects have to be seen as possessing instinctive, self-determinative purpose. This purpose is restricted by various conditions such as temperature, light, humidity, magnetism, and atmospheric pressure, but we need to be clear about the self-determination involved in the freedom to change within in a fixed, closed field.

The problem of determinism and free will has been debated throughout the history of philosophy and ethics and intertwines with the question of the determination of a field and variability within it. Determinism is suited for explaining the determination of a field, but when it comes to choice among variables, we are not mistaken in speaking of free will.

In the field that Loeb set up, certain living things show negative phototropism and others positive phototropism, which means that each of the life forms was brought up within a different field. If their distinctive fields were taken away, those life forms would shift to free activity. This means that we have to recognize close ties between self-determining, purposive behavior, and the determination of the field. The Buddha, Confucius, and Jesus all went to sleep at night, entering the animal state of negative heliotropism. When the sun rose, they woke up, demonstrating their heliotropism. They acted freely under the sun, that is, they were able to make choices to adapt to variations within a certain determined field. In this sense, not even the free will of these saints and sages was detached from the determination of a field. In addition, there is no need to restrict this field to the space within which living organisms are active. We may also consider the field of the living organism itself, which enables ample freedom of activity. For example, the human body temperature of 37°C can function in an outside temperature of between 1° and 27°C, even if it is naked, wrestling or running a race.

Theologically, Calvinistic determination and Arminian free will theory can be harmonized if seen as the free-will of variation within a field.

Mechanical conditions of freedom

Self-determining purpose is not determination by chance. From a large number of variables, spatial as well as temporal, the conditions most suited to oneself have to be chosen. Thus the more severe the variation, the greater the need for making a selection. For this reason, the concentration of attention and the accumulation of past experiences—that is, the systematic constitution of memories, associations, judgments, inferences, and the like—are steps leading to conviction. Beginning with the powers of attention, the steps that culminate in conviction must be organized as admirably as the brain is or they cannot possible synthesize and regulate the process of selection. Experiences recorded in the tens and hundreds of thousands have to be classified and connected rationally in precise mechanical operations for the goal to be reached. The achievement of self-determining purpose only becomes possible through the mechanical operations of cerebral structures.

Finality and the structure of the brain

Ernst Scharrer's work on the structure and functions of the brain brings together recent research on purposiveness in the brain. He begins with the example of invertebrate hydra, whose nerve cells and protuberances run through the body like the mesh of a net, enabling sensitivity to stimuli (light and touch) the animal needs and bringing about reactions in response to a goal (the capture of prey and the like).[16] According to the

---

16. Ernst Scharrer, *Vom Bau und Leben des Gehirns* (Berlin: Springer, 1936), 9.

calculations of Constantin von Economo, there are 14 billion cells in the cerebral cortex of human beings.[17] The cerebellum is the center for purposive dexterity in the limbs,[18] and the collaborative task of preserving harmony takes place in the realm of the axial nerves. Only when this regulation has been achieved can human activity begin, which explains why infants cannot walk.

> These various reports are organized in the medulla oblongata where the code is deciphered and translated into the language of the axial nerves and transmitted to the cerebellum, so that taking a measurement with their eyes, humans can walk towards an *object*.[19]

The unconscious act of *reaction* is also seen to be a purposive response without any connection to the will. Contrary to theories that see the brain as a simple machine, due attention has always to be given to the finality of the brain.

Scharrer further points out that when the spinal cord is lower than the spinal column, so that the spinal column that should be attached to the greater part of the spinal cord comes up short, development is arrested. But he also reports that where dye pigments and medications fail to reach the brain, the brain's functions preserve themselves.[20]

Selective purpose in human sight

D. H. Martin has studied the absorption of various materials within the eye. His results shows that with light longer than 295 $\mu\mu$, absorption in a transparent corneal lens begins from 400 $\mu\mu$, and that at 350 $\mu\mu$ everything is absorbed. Therefore, in the case of light less than 300 $\mu\mu$, those that are less than 400 $\mu\mu$ probably cause a chemical reaction within the crystalline lens of the cornea. Vitreous humor has an absorption band of between 280 and 250 $\mu\mu$, but it is not yet understood what this means.

Vitreous humor and aqueous humor allow passage from around 300 $\mu\mu$, but the crystalline lens shows strong fluorescence with light from 360 to 400 $\mu\mu$. This being the case, most ultraviolet rays do not reach the retina, which means that inflammation caused by ultraviolet rays is restricted to the outer part of the eye.

17. *Ibid.*, 39.
18. *Ibid.*, 43.
19. *Ibid.*, 71–2.
20. *Ibid.*, 88.

Knowledge of Cosmic Purpose

When we look at material objects and perceive light and dark, some sort of spectrochemical or photoelectric change has to occur for sight to take place. Already since 1851, something called rhodopsin or visual purple has been extracted from frog eyes. This pigment is known to be present also in the eyes of humans, monkeys, livestock, doves, and other animals. When light is added to this material, it loses its crimson-purple hue and turns brown. Moreover, according to Eldridge Green, the mechanism of sight is a spectrochemical reaction of photosensitive matter:

$$\text{rhodopsin} + \text{light} \rightarrow \text{degraded product (colorless)}$$

This degraded product works in the system of light nerves to transmit a stimulus to the brain. Rhodopsin is said to be regenerated in the retina to prepare for the light to follow.

We understand from the experiments of Selig Hecht that when exposed to light, the suction tips of clams move 0.2 to 1.5 centimeters in the direction of the shell, and that this always occurs about two seconds after the light has been dispersed. Changes in reaction time are shown to be precise measurements based on the strength and length of the light waves. 500 $\mu\mu$ is the most effective, but no sensation at all occurs when the light either decreases to 420 or increases to 620 $\mu\mu$.

The type of green algae known as volvox globator also moves towards a light source. Its most effective length is 500 $\mu\mu$ and loses its effect on both sides as it is distanced from 500 $\mu\mu$, until at 420 and 620 $\mu\mu$ it becomes a flat zero.

These spectrochemical reactions are extremely efficient and, at first glance, seem simpler than other biological reactions.

Eyesight is thought to be very closely related to changes in photosensitivity in such cases, but a pure method of retrieving the data remains largely impossible at this time.[21]

---

21. Ariga Teru 有賀輝,『光化学』[Spectrochemistry] (Tokyo: Sankaidō, 1930), 338.

8 The Emergence of Self-Conscious Purpose

The unfolding of self-conscious purpose

Consciousness and the brain

The brain is a complete machine, but not a machine that immediately becomes conscious. Only when life runs through it and is given self-determining purpose does consciousness become connected to the world of life. Of course, there are cases when consciousness is present without actualizing self-determining purpose, as in dreams. Observations, associations, and memories reappear in the dream world, besides which judgment and reasoning can also occur. But intentionality in the dream world is completely removed from the real world.

How does this subconscious come about? We may speak of it as a special instinct in the mechanics of the brain whose self-determining purpose would be absolutely impossible without this dream-like mechanism.

The six mechanisms appearing in the natural world—energy, change, growth, selection, law, and purpose—are reconstituted in the realm of life where the microcosm of "consciousness" is constructed.

1. The first stage in bringing the objective world to subjective consciousness is *attentiveness*. The power of attention transmits all of one's experiences to a "living" storage battery made up of some 14 billion brain cells. The battery analogy may not be quite right, but at as a metaphor it is useful. I realize that something said to be *alive* is different from a storage battery, but I also see that life itself functions by means of electricity. Still, it is obvious that *electricity* and *life* are not the same thing. As I have pointed out frequently, life is like a flame ignited by a combination of all the laws that run through the structure of matter. Through the use of the design purpose in electricity, the impulsive purpose of life is granted attendance.

2. Attentiveness leaves an impression. 14 billions cells *live* within the

brain like so many photographic plates. The work of *association* connects the multiple forms of living photographic plates that work like storage batteries. When association is needed, it may be that tens of "filaments" in the cells are connected together, much the same as what happens when the voltage is raised by linking several batteries together. The strength of association lies in the speed with which these connections are made.

3. Change is unified horizontally with this power of association, but the function that aligns them vertically and according to temporal transition in a progressive and well-regulated fashion belongs to *memory*. I would hold that memory and attentiveness bring about a temporal (vertical) joining of units of consciousness preserved in the storage batteries of the photographic plates of brain cells.

4. The sifting of the horizontal and vertical world—namely the world of space and time—provided by association and memory and its classification into right and wrong, true and false, beauty and ugliness, is the work of *judgment*.

5. The discovery, through judgment, of a fixed arrangement and order in the objective world (including the objectification of the subject) belongs to the power of *reason*.

6. *Intentionality* occurs by combining the purposiveness of attention, association, memory, judgement, and reason and making use of them all. Only then can self-conscious purpose be carried out.

Self-conscious purpose and the realm of "I"

As stated above, retrieval of the objective world takes the form of a microcosm, namely the realm of consciousness. Consciousness itself is sensitive to the impulses of life that have an immense influence over the intentionality of subjective feelings of excitement, tension, fear, anger, passion, desire, and the like.

But even in such cases, lower feelings can be sifted out and self-conscious purpose achieved through the action of the natural abilities of selection and reasoning.

We may be permitted the remark that the achievement of self-conscious purpose in humans is rather more complete than what we see in other life forms. As I stated earlier, in lower life forms, even if a capacity for bringing about self-determining purpose is present, either the range of experience is so narrow that the achievement of an adaptability requiring almost no development of brain functions suffices (as in the reproductive activity of kelp and ferns); or they spend their lives in activities ordered by progressive purpose (as in the case of bacterial enzymes). In neither case do they have the kind of high-level nerve centers we see in insects.

The need for some kind of complete mechanical cerebrospinal system increases when we come to the higher animals, given their drive to expand the reach of self-conscious purpose through rational discovery and invention, to conquer various environments without misjudging changes and transitions.

The resulting living mechanical unity is given to humanity as consciousness, the engagement in a mysterious psychological life not readily found in other animals that enables them to see dreams and have visions.

Once again, we need to remember that the machine is not an *I*. The brain is part of a necessary process for carrying out self-determining purpose. Without that process, there is no *I*. The *I* represents a composite purpose that links one purpose to another. It is a simple purpose, the primary purpose. In other words, *I* means *self-determination*. The individuality of the *I* consists of a synthesis of self-determining experiences.

Even the apparently mechanical nature of conscious processes themselves is a gift in the form of a crystallization of self-determination. If attention lacked self-determination, the images it leaves behind would not be able to recall associations and memories or awaken judgment and reason. In the end, there would be no invention and discovery. Without self-determination and its expansion of the reach of self-conscious purpose, invention and discovery are completely absent.

Those who take only the machinery into account lean towards determinism; those who see purpose, lean towards free will theories. I myself hold that machines exist to realize purpose, that there is no purpose without machinery and no machinery without purpose.

The day the brain breaks down is the day consciousness vanishes from the face of the earth. But even should the machinery be destroyed, we cannot suppose that the laws of the universe that realized the *I* would perish. With only materialistic theory, it is difficult in the extreme to explain the genetic transmission of musical talent from parents to children and to grandchildren. Unless we rethink the world of laws as a prearranged finality, the question of how a talent for music is woven into minute genes will remain unanswered.

Finality gives us the *I*. Along with the finality that has come about in the cosmos, there is an *I* preserved in that finality. The fundamental

The Emergence of Self-Conscious Purpose

reality of the cosmos that has given me memory surely remembers me for all eternity.

Modern psychology and the question of consciousness

The nervous system developed at first as a center of movement for animals. As University of Chicago neurologist C. J. Herrick has noted, the progression to a center for receiving environmental stimuli belongs to a later period, even in the case of higher animals.[1]

Scientifically speaking, it is worth noting that the story of the development of modern psychology went through this very same process, as R. S. Woodworth's study bears out.[2]

In 1879 Wilhelm Wundt set up the world's first laboratory for experimental psychology in the University of Leipzig. Across Europe and the United States, "functional" and "structural" schools came to birth. In 1912 John B. Watson, professor of psychology at Johns Hopkins University devised a "behavioral" approach to psychology as a reference point for modern psychology, based exclusively on objectively measurable animal psychology. The publication of his first book in 1914, *Behavior*, caused a stir in the United States, witnessed by his election to the presidency of the American Psychological Association the following year.[3]

In 1914 I studied experimental psychology at Princeton, where Watson's book was highly regarded. Having just left Japan after publishing a book called *Studies in the Psychology of the Poor*, I felt hemmed in by the overly narrow scope of Watson's psychology.

William James, the founder of experimental psychology in the United States, had included consciousness in the study of the psyche. The behaviorists rejected the idea, but their contributions to human psychology were considerable and are understandable from the focus on the behavior of the nervous system in lower animals that they took as a starting point.

We must acknowledge the great strides made by behavioral and work psychology in the study of child development, industrial psychology, educational psychology and in particular mentally disabled children, criminal psychology, therapy for the mentally ill, and so forth.

---

1. Charles Judson Herrick, *Neurological Foundations of Animal Behavior* (New York: Henry Holt, 1924).

2. Robert S. Woodworth, *Contemporary Schools of Psychology* (London: Meuthen, 1951).

3. *Ibid.*, 70.

At the same time, it is regrettable that so much weight given to objective measurement resulted in a neglect of one important aspect of human psychology, namely introspection.

Modern physics has also to recognize, as Heisenberg's uncertainty principle has made clear, that when we enter the minute, microscopic world of Planck's quantum mechanics, observation is altogether impossible. When a change in operational procedure occurs, actual facts are destroyed. This does not mean we then deny the existence of a microscopic world. The use of stochastics, as in the application of "statistical probability" in physics, cannot be dismissed as unscientific. Watson rejects this, and in doing so, leads behavioral psychology to a dead end. For him, the sole basis for psychological activity is one of response to an external stimulus: S–R. Like an inorganic chemist compounding the various elements through their atomic values, Watson tried to make S–R the basic elements of psychology without giving a thought to the elements of inner reflection and introspection.

The Gestalt school in Germany emphasized sensation through organization. Its pattern of stimulus–organization–response did not ignore subjectivity like the behaviorists before them.

After completing medical school in England, William McDougall turned to the study of social psychology and stressed consciousness in the sense of the knowing subject.[4]

McDougall recognized the emergence of purpose through consciousness, and his treatment of the question from the standpoint of social psychology represents an important chapter in the history of modern psychology. We may summarize his points as follows:

1. *A certain activity is forced.* Behavior probably begins with a stimulus, but even if the stimulus is taken away the behavior continues.
2. *Change occurs in the forced activity.* For example, when a sound is made, a rabbit scampers into its hole.
3. *When the goal is achieved, the behavior stops.* After this, other behavior begins. A cat runs to a tree and climbs up the trunk. It comes to rest on a branch and looks down quietly at the dog.
4. *Behavior makes progress through repetition.* As non-beneficial activity is halted and manners become refined, the goal is reached more efficiently.

In this way McDougall restores to purpose the emphasis that Watson had

---

4. *Ibid.*, 217.

denied it. This same tendency gradually showed up among Watson's followers. Clark L. Hull, among others, came to acknowledge the "purposiveness" that Watson had overlooked.[5] While he did not deny the existence of consciousness, he considered it a mere phenomenon.

Even within Watson's school, some like Edward Chace Tolman allowed for purposive activity,[6] though he did refrain from treating the problem of consciousness.

The importance of consciousness among religious and social psychologists

While behaviorists were afraid to even touch on purpose and consciousness, in the effort to treat the human being as a whole, some scholars turned to the psychology of religion under the influence of William James. Social psychologists, meantime, tried to deal with the social nature of human beings. Dissatisfied with the movement that set out from Watson's S–R, they moved into the deep, inner recesses of human psychology. As if in agreement, the Freudian school forged ahead from nervous disorders to pursue the psychoanalysis of the unconscious and subconscious mind, opening a field the behaviorists has closed off.

Freud himself tried to explain the super-ego from a psychoanalytic point of view. Moreover the Duke University psychologist J. B. Rhine pioneered the field of parapsychology, which the Japanese psychologist Fukurai Tomokichi had touched on forty years previously.

In today's age of television, it may not seem so remarkable, but statistical calculations of the reliability of laboratory experiments on things like clairvoyance and mind-reading yielded results of high academic interest.

James broke new ground in the psychology of religion by studying the life stories of people based on their "whole lives." I also have to admire the charming way that Rhine analyzed the "super-ego" statistically and tried to draw it into the realm of probability theory. Religious cultures East and West are full of material about such things, and I consider it a sign of negligence that only psychology has taken up such questions so far.

Nerve fibers and finality

The neurological specialist Katō Fusajirō told me how completely surprised he was to see a movie showing nerve cells at work "like a carpenter

5. *Ibid.*, 69.
6. *Ibid.*, 103, 105.

building a house." That was June of 1957. I contacted the producer of the movie, Nakazawa Tsuneyuki of Keiō University's Faculty of Medicine, and he agreed to let me see it.[7] Chicken eggs were warmed in an incubator and on the twelfth day the brains of the chicks were removed from within the eggs and dissected, and the development of their nerve fibers captured on film. I imagined that the nerve cells would have no inclination of their own and nothing to guide them. But what surprised me about the movie was the step by step guidance involved in collecting forty or so neuromuscular fibers into a single strand, and in the Schwann cells that would serve as a "sheath."

There are three parts in the glia that become peripheral nerves: (1) macroglia, (2) microglia, and (3) neuroglia or oligodendroglia. The lively activity of the oligodendroglia was clearly visible in the film. It was as if one were watching a sage at work, pulling the nerve fibers out one by one to the end, and then returning to the start, until they were all gathered into a single strand. This display of what we may call intentionality reminds us altogether of the finality of the "organizers" that the embryologist Spemann found to play an important role in nerves. Here again I was directed to the realm of finality.

Nerve fibers connect the brain with peripheral nerves. As Ernst Scharrer has noted, during their passage from the brain and around the spinal cord to the point of contact, there is no break at all in continuity. Nothing can compare to nerve fibers whose long, thin cells work like a single electric wire with no short circuiting. Just seeing this in the film, I could not restrain my wonderment at the adaptability of physiological mechanisms in the universe.

Darwin's teleology

Darwin on emotional expression in animals and humans

Darwin denied that the origin of the species derived from finality in natural selection. But as he carried on his work in the life world, he found himself obliged to acknowledge finality to a certain extent. We see this in his study on worms and soil,[8] but still more clearly in his book on the expression of emotion in animals and humans. In the first six chap-

---

7. See above, pages 116–17.

8. *The Formation of Vegetable Mould through the Action of Worms, with Observations on their Habits* (London: John Murray, 1881).

ters Darwin affirms that a degree of finality appears in the emotional expression of animals.

This is understandable. On one hand, the snakes best at threatening enemies into running away are those that best avoid becoming the prey of their enemies. On the other hand, the enemies of snakes best adapted to the dangerous work of killing and eating poisonous snakes are the most numerous to survive. Darwin notes that:

> in the one case as in the other, beneficial variations, supposing the characters in question to vary, would commonly have been preserved through the survival of the fittest.

On this basis alone, it is difficult to understand why unique changes occurring in individuals involve finality. A few lines earlier, he remarks:

> We have also seen that some porcupines, when angered or alarmed at the sight of a snake, rapidly vibrate their tails, thus producing a peculiar sound by the striking together of the hollow quills. So that here both the attackers and the attacked endeavour to make themselves as dreadful as possible to each other; and both possess for this purpose specialised means, which, oddly enough, are nearly the same in some of these cases.[9]

Taken together, these two ideas show Darwin distancing himself from the views of *Origin of Species*.

1. *Confirmation of purpose*. Given that Darwin did not believe in discontinuous evolution, the attempt of *Origin of Species* to see continuous change everywhere left no leeway for incorporating finality, which implies discontinuity. But from the time of his study of earth worms and soil, he was no longer able to avoid recognizing finality, so that when he came to emotional expression in animals and men, circumstances obliged him to all but take up finality full-scale.

2. *The changes of finality precede natural selection*. In *Origin of Species* change occurs by chance, and this is part of a way of thinking organized around natural selection. In his work on emotional expression, the casualist approach yielded to the acknowledgement of purposive change.

By themselves, the two remarks cited above do not amount to an affirmation of finality, but let us look at how teleology appears in other chapters of the book.

---

9. *The Expression of Emotion in Man and Animals* (New York: D. Appleton, 1872), 110.

Cosmic Purpose

In the conclusion to his study of emotional expression, Darwin avoids speaking of purpose, but in the opening chapter he cites Eduard Pflüger to express full support for the presence of purpose and harmony in the muscle activity of frogs.[10]

In addition, he agrees completely with finality in retinal contraction:

> A radiation of nerve-force..., as seen in the contraction of the iris preventing too much light from falling on the retina, might afterwards have been taken advantage of and modified for this special *purpose*.

He goes on to say the same of reflex actions:

> Thus reflex actions, when once gained for one *purpose*, might afterwards be modified independently of the will or habit, so as to serve for some distinct *purpose*.

Turning to continuous habitual movements in lower animals, he writes:

> I have already given in the case of Man several instances of movements associated with various states of the mind or body, which are now purposeless, but which were originally of use, and are still of use under certain circumstances. As this subject is very important for us, I will here give a considerable number of analogous facts, with reference to animals; although many of them are of a very trifling nature. My object is to show that certain movements were originally performed for a *definite end*, and that, under nearly the same circumstances, they are still pertinaciously performed through habit when not of the least use.[11]

Later, affirming finality in activities related to emotions, he acknowledges that nerve movement sometimes occurs without any direction.[12]

In the fourth chapter of the book, Darwin gives a clear statement of the fact of finality:

> Involuntary and purposeless contractions of the muscles of the chest and glottis, excited in the above manner, may have first given rise to the emission of vocal sounds. But the voice is now largely used by many animals for various purposes....

He then goes on to repeat his affirmations of finality in animal voices.[13]

Regarding purpose in involuntary muscle movement Darwin argues:

---

10. *Ibid.*, 36.
11. *Ibid.*, 41–2; emphasis added.
12. *Ibid.*, 51, 65.
13. *Ibid.*, 84, 103.

> In this manner actions performed by the contraction of voluntary muscles might have been combined for the same special purpose with those effected by involuntary muscles. It is even possible that animals, when excited and dimly conscious of some change in the state of their hair, might act on it by repeated exertions of their attention and will; for we have reason to believe that the will is able to influence in an obscure manner the action of some unstriped or involuntary muscles, as in the period of the peristaltic movements of the intestines, and in the contraction of the bladder. Nor must we overlook the part which variation and natural selection may have played.[14]

In this way Darwin gives full assent to the idea that the psychology of animal emotions comes about through finality. It is curious to observe how the casualist explanations embraced in *Origin of Species* are shut out here:

> We have, also, seen… various voluntary movements by threatening gestures, opening the mouth, uncovering the teeth, spreading out of the wings and tail by birds, and by the utterance of harsh sounds; and the purpose of these voluntary movements is unmistakable. Therefore it seems hardly credible that the coordinated erection of the dermal appendages, by which the animal is made to appear larger and more terrible to its enemies or rivals, should be *altogether an incidental and purposeless result* of the disturbance of the sensorium. This seems almost as incredible as that the erection by the hedgehog of its spines, or of the quills by the porcupine, or of the ornamental plumes by many birds during their courtship, should all be purposeless actions.[15]

All of this shows how, when it came to questions of psychology, someone with a deliberately anti-teleological stance was unable to sustain the arguments that ran through *Origin of Species*, finding himself forced to confess he had been driven to a recognition of finality within chance.

THE UNFOLDING OF CONSCIOUS PURPOSE

A short history of teleology

In the history of modern philosophy, the conscious construction of teleological theory begins with Kant's *Critique of Practical Reason*. Up until then, teleology had followed in the line of Aristotle's "organic

---

14. *Ibid.*, 103–4.
15. *Ibid.*, 102.

teleology" as expounded theologically in the scholastic philosophy of Thomas Aquinas. Not even the leading Protestant theologian John Calvin advanced beyond Aquinas. This is why "teleology" has had such a poor reception among scientists in modern times.

With the arrival of quantum mechanics in modern physics, the attitude to teleology underwent a rapid change. Sommerfeld introduced "selection rules" into spectrochemistry and took a public stance acknowledging teleology in crystallography. His viewpoint differs considerably from that of the mechanical physics of Kant's time.

Advances in chemistry gave further impetus to this tendency. Oparin affirmed a "teleology of matter" from a materialistic standpoint. Research in enzymes has made his claim obligatory today.

Nevertheless, a finality able clearly to grasp the development of the three levels of motivation, means, and ends with conscious content, such as we find it in Kant's *Critique of Practical Reason*, could only offer faint proof through the affirmation of mental states. Even H. G. Wells recognized a "teleology" in the realm of human mental behavior.[16]

Even where purposiveness is consciously held to be true, humanity has not given it proper attention. Purposiveness used for destructive ends or toyed with playfully differs little from unconscious selective activity.

If all of humanity were able to bring purposiveness to full consciousness, they would be able to work together towards world peace, to divert the energy spent on war to organized movements for global cooperation, and then turn their remaining strength to discovery and invention.

The civilizations of ancient Greece and Rome collapsed from internal strife and invasions from without. In the same way, if the atomic energy human beings have invented is applied to the destruction of human life, I fear that the time will come when, like the reptiles at the end of the Mesozoic period, our marvelous modern scientific civilization will also be driven to the brink of destruction.

The history of trigonometric functions

As a perusal of the history of science will show, the development of human science covers a mere two thousand and some hundred years. Reading the history of mathematics, we find that the awakening to trigonometric functions took place in the scientific culture centered in Ephesus in Asia

---

16. Book 8, Chapter 7 of *The Science of Life* (see page 39, note 2 above).

The Emergence of Self-Conscious Purpose

Minor. Thales, one of the seven sages of Greece, was raised on the Island of Miletus near Ephesus. He is credited with inventing trigonometry. The Sicilian scientist Apollonius made extraordinary advances in astronomy. But it was only as a result of developments of mathematics around the middle of the ninth century in Baghdad, the center of Mohammedan culture, that trigonometric functions proved beneficial to astronomy and navigation through the use of natural logarithms.

Mathematics and the logic of finality

When it comes to the establishment of conscious purpose, nothing is as clearly instructive as the development of mathematics. Trigonometry became possible only with advances in arithmetic, geometry, and algebra. The three choices trigonometry offers made it easy to calculate the unknown quantity of an apex from two angles and a base.

To achieve this goal of simplifying complex calculations, previously developed methods had to be bracketed, exponents used and displayed in logarithmic tables, calculations abbreviated, and laws of selection drawn on to allow for new numbers to be inserted and removed. To simplify the whole process, a large number of rules had to be arranged in as simple a form as possible, collecting genuinely complex calculations into differential and integral formulas. Only when the six elements of what I call "the logic of finality"—energy, change, selection, process (growth), law, and purpose—are lined up, can purposiveness come about.

The history of modern mathematics may be called a history of the emergence of conscious purposiveness.

In Kepler's age, the conic curve discovered by Apollonius of Perga in Asia Minor was thought to be no more than a mathematical game. People were said to have been surprised to learn that it could be used in formulas for the movement of celestial bodies.[17]

A priori selection in the fourth dimension (time)

The third a priori selection expressed in the form of "life" is selection with high-level concentration, accumulation, unity, and planning. Developed to the fourth dimension of time, it manifests the highest level of a priori selection, one which opens our eyes to the consciousness we call "mind." This level of selection calls forth the appearance of a unitive conscious-

---

17. Kobori Akira 小堀憲, 『数学史鈔』 [Extracts from the history of mathematics] (Osaka: Akitaya, 1946), 15.

ness wherein "the whole world is centered in the self." One suddenly perceives that "I am I," "I am given life," "I have been given a mind." One becomes aware of how strange and surprising the "self" really is.

From all of this we understand that the realm of purpose rises to consciousness and that "matter" comes to consciousness as "purpose." Here the world constructed by the materialists is turned on its head and made conscious. In the end, we establish that the cosmos is not constituted of matter resulting from chance, but from selectivity.

Finite awareness of being given life as an *I* with a *mind*, that is the awareness of a priori selectivity, contains five types of elements:

1. The intuition of a priori selectivity through a priori selection. This means to come to the ultimate a priori selection which, in self-consciousness, grasps the being of "self" in the "objective case."
2. The self-awareness of a finite, relative *I* given through a priori selection.
3. The self-consciousness of a subjective *I*, given through a priori selection, with the power to preside over the constitution of a microcosm through response to the objective world by means of the a priori capacity for sensation, instinct, and consciousness.
4. The intuition of categories of space and time given through a priori selection. Human beings do not intuit the a priori selectivity of the atoms and molecules that make the earth turn and organize the self, but they do possess the intuition of a world formed as space, together with the intuition of energy in constant flux and its irreversible directional selectivity in the form of time. All knowledge of space and time begins from the self-awareness of selectivity.
5. The self-consciousness of the bridge to a posteriori selection dependent on a priori selection and its development—namely, the self-consciousness given through a priori selection—devises a bridge to the a posteriori selectivity of the powers of attentiveness, memory, association, judgment, reasoning, aesthetic appreciation, emotion, and so forth. By developing in this direction, self-consciousness is able to analyze a priori and a posteriori selection completely.

The forward advance of purpose

A priori selection passes through five stages before arriving finally at the constitution of an *I*. The *I* itself is a *subject* that is itself an *object* of knowledge with *purposiveness*. A priori selection that has succeeded in bringing about the *I* may thus far be said to manifest a selectivity aimed at the creation of an *I* grounded in a degree of freedom latent in life.

Apart from the sphere of a priori selection in its primary to its tertiary forms, and through countless combinations and displays of its innate abilities, the *I* opens up all sorts of activities with a posteriori purpose, finally breaking through to a new creation. This is what leads to the birth of a world of new inventions, discoveries, arts, and values.

The special characteristic of selective teleology

Aristotle strived for an "organic teleology." Kant stressed a fourfold teleology: organic, logical, moral, and aesthetic. Darwin insisted on blind natural selection and Ernst Troeltsch advocated a historical teleology. I hold that nature has "selective purpose." My aim is to point out that the selectivity flowing beneath the three types of purpose—organic, tendency to life, and conscious purpose (logical, moral, and aesthetic)—possesses the constitutive elements of purpose.

The reason we need to include the development of latent purpose is that hidden within the developmental process of selectivity is an organic life that shows up in the normativity, maintenance, balance, mechanism, unity, aggregation, causality, orientation, and tendency of selectivity.

We know that there is a strange and wondrous law of conservation in the essence of the universe: in the world of energy, light waves expand and, in order to prevent the consumption of energy, undergo a creative transition to elementary particles as "light" by transforming wave movement into circular movement, taking a form like the conservation of matter; heat waves appear at the tail end of entropy, and just before energy vanishes, are constructed in the form of "life." Then, at the end of life's developmental process, mind appears, and with it the surprising creativity, preservation, and recreation of consciousness. The social and cosmic expansion of consciousness with a deep degree of solidarity is a fact that merits careful attention.

In this way, the structure of the selection principle points to the essence of the universe, with the reality of the selection principle determining the reality of cosmic purpose itself.

The emergence of purposive consciousness

When a living organism takes in the outer world through the senses, it has to adjust to solids, liquids, gases, light, and radiation through the sense organs with which it has been endowed. All of this has to be seen as serving the mechanisms of adaptation in the universe.

For solids, the function of touch develops; for liquids, taste; for gases, hearing; for light, sight; and for radioactivity, smell.

From the lower animals to the higher, the development of the sense organs and their location differs from one life form to the next. The ears of insects are connected to the upper part of the lower extremities; the eyes of snails protrude at the front like antennae.

But sensation also developed by adjusting to abnormalities in living organisms, a fact that is hard to comprehend without an underlying cosmic purposiveness.

Besides the five sense organs, there also developed sensitivity to things like (1) heat, (2) cold, (3) motion (including things like dizziness), (4) pain, (5) pressure (including atmospheric pressure), (6) itching, (7) tickling, (8) humidity, and (9) carnal pleasure.

The range of sensation differs considerably from animal to animal. Sensitivity to ultraviolet light is vastly different in the vision of insects from what it is in human vision. The same can be said of smell. Certain insects can pick up the scent of a female from several kilometers away, as the entomologist Jean-Henri Fabre has noted from his experiments.

Human nerves are said to constitute an electrical system, but unlike the long-range we expect of electrical power, that power is sluggish in human nerves, coursing at only about 200 meters a second.

If human beings perceived the spinning of the earth and if they were sensitive to all electric wave lengths, they would surely live hectic lives and be forever on edge. It is probably due to the dullness of their nerves that they can lead short but relaxed lives. We should probably see this also as adaptability to the natural world.

We will leave the physiological analysis of the sense organs to the specialists and look at their relationship to mental mechanisms.

Hermann von Helmholtz has criticized the human sense organs for serious imperfections. Without denying these imperfections, we should not forget how mental capacities step in to compensate. The human eye, like a telescope, has been made with good distance vision, but how about when it is using a microscope? Given its ability to employ all manner of optical devices, does not the simple and imperfectly constructed human eye actually appear to be rather well made? St. Paul writes that "my (Christ's) power is made perfect in weakness" (2 Cor 12:9). The same paradox holds true for human sensation.

The marvel of sight is that, although the human eye should see things inverted, as they do when viewed through an old-fashioned camera, they are not transmitted in this form to the brain but are seen as perfectly upright. How are we to explain this marvel? Jesus commands us to "be perfect as your heavenly father is perfect" (Matt 5:48). When Nietzsche conceived of the consciousness of the "Overhuman," he explained it through the creative consciousness of Jesus the carpenter.

Human consciousness is not simple. Memory, association, judgment, reasoning, learning, experience, imagination, attention, conviction, feelings, courage, sympathy—all knowledge, emotion, and volition rises to consciousness in a synthesis.

Let there be no mistake: consciousness is a product of *purpose*. Knowledge is made possible by purposive, psychological selection. Through knowledge memory is born, and this memory is a non-spatial, temporal representation of *purpose*. The non-spatial representations of memory become association; and the assemblage of these non-spatial associations becomes "imagination," without which there would be no "invention." This non-spatial "opening up" enables a reconstitution of our views of the universe and the birth of personal cosmic consciousness.

The disabled American genius Helen Keller, like Emanuel Swedenborg (1688–1772) before her, has embraced mystical philosophy. She is unable to see, to hear, or to speak, and yet she expresses philosophical ideas freely. She has *received* an idea of the essence of the universe. This idea is no mere subjective actualization; it is perceived as an objective actuality.

Even things that cannot be perceived sensibly in the outer world can be perceived through different operations of the sense organs. These operations allow for things received subjectively to be expressed in objects and reported to the outer world. When these reports are confirmed to be without error, the gap between subject and object is removed.

The unification of initial purpose and ultimate purpose in consciousness

Human activity displays both motivations and ideals. Motivations are the initial purpose and ideals the final purpose.

In the culture of modern physical chemistry, the complex measures of experimentation trail the interval between the initial and final stages like a shadow. Thomas A. Edison came up with the ideal of inventing the light bulb, and with a grant of $40,000 secluded himself in the woods of West Orange, New Jersey. Even after five years and 110,000 failed experiments, he kept on working until finally, in October of 1878, his experiments succeeded. For modern civilization this is a more momentous event than

Napoleon's crossing of the Alps to conquer Italy. Edison linked the facets of motivation and ideal through experimental trials.

The invention of the airplane, the wireless telephone, and the television all followed the same method. In short, the knowledge of the laws of physical chemistry and their combination progressed purposively, one step at a time, with an eye on the final goal. Just as the energy levels of electrons within the atom have fixed, integral norms, invention needs to go through trial after trial until it discovers a certain point of stabilization. Invention represents the point at which inference and existence become one.

The establishment of centers for scientific research is proceeding at a rapid pace around the civilized world. But no matter how many centers there are and how fantastic the inventions, there is no ignoring the laws of physics, chemistry, and biology.

Many inventions, therefore, are born of specialists paying attention to experiments. That young, uneducated workers often come up with inventions is due to specialized knowledge. New inventions are born from the university classroom because of the many opportunities provided there for discovering a new axis to spin their ideas around.

Many of the new discoveries and inventions made by young students are due to their ability to look at things from a different angle. Birds had wings before humans invented the airplane. Fireflies were equipped with luminescence before humans invented a way to generate light. If human beings could know the myriad laws of the universe in detail, the wonders of their inventions would surpass imagining.

In terms of inventions, we may be said to live in "an age of rapid transformation." But if human beings give in to a reckless use of the atomic bomb, the ensuing degeneration of the human race will leave behind the age of rapid transformation and civilization will return to a savage age.

If such days were to come upon us, cosmic consciousness would be lost and we would fall subject to animal instinct.

9 Cosmic Evil and its Salvation

THE FINAL END OF CREATIVE COSMIC EVOLUTION

The natural sciences teach us details about the mechanisms in the universe for evolution. Prout's hypothesis of the evolution of the atom has received the support of atomic physicists.

Theories of plant evolution have also thoughtfully instructed us concerning the botanical lineage from seaweed to flowering plants.

Scientists have been conscientious in relating to us the story of the evolution of the animal world, as the record of advances in paleontological history, comparative anatomy, ontogeny, embryology, and genetics all attest. If there are reasons to believe that evolution moves forward, where does it finally end?

I do not see any need to think about this question statically or in terms of cause and effect. If we suppose that the evolution of the universe has advanced orthogenetically in the direction of human spiritual evolution, it only seems fitting that we approach the development of the insect world analogically.

If we divide ants into the three classes of queen, males, and workers, the workers among leafcutter ants demonstrate six types of physiological division of labor.[1]

Some ants specialize in cutting leaves, other worker ants are specialists in protection, and still others have the task of nurturing the young hatched from the eggs. Such physiological differentiation is not present in humans, but at a psychological level, educational systems allow for hundreds and thousands of divisions of labor. Human societies can subsequently enhance the development of this psychological differentiation all the more.

---

1. See William Wheeler, *Ants: Their Structure, Development, and Behavior* (New York: Columbia University Press, 1910).

Scientific progress has slowly increased the need for the division of labor. No one human brain can handle everything; it will take the organic cooperations of billions of superior brains if we are to plan the evolution of humanity. The sciences dealing with the evolution of sight, hearing, touch, smell, and taste, the science of cooperative societies, as well as measures to promote the evolution of human sensation and will as they relate to beauty and goodness may all be planned.

The advance of civilizations and cultures made possible by the ordered unity of billions of brains will proceed at an unimaginably quicker pace. The end result has to be the realization of a creative world of freedom. If humanity can be patient in the meantime, our common mission will be all the greater. If we do not plan for its realization, and if the factors involving the physiological, psychological, and social decline of humankind are allowed to multiply among living organisms, the orientation of cosmic evolution will probably take a strange direction.

Even if the atomic bomb should drive humanity to the brink of destruction, we may suppose that the evolution of the atom, of the earth, of vegetation will lead to an evolution of animal life that excludes humanity. By eliminating the all too contentious and materialistic human presence, we may suppose that the evolution of races of chimpanzees and the like will take the place of human beings like insects teeming on the earth, in the water, and in the sky.

The lungfish in Australia tried to climb out of the sea directly on to land, but the inner system of their bodies was not suited to life on land. Their skeletons were weak and other internal organs relatively undeveloped. Moreover, their outer layers of skin were not tough enough for survival in the atmosphere. So they returned to the sea and developed into bony teleosts.

If we examine the history of the evolution of organic life, we find that, as a rule, those that veered too far from the path suffered regression. What is lamentable about humanity is that we have wasted too much energy on hostile behavior.

Like the overly belligerent dinosaurs that became extinct around the end of the Mesozoic period, has not the human race with its invention of the atomic bomb been hastening the arrival of its own self-extinction?

The author of the book of the Apocalypse writes of the death of all the animals in the sea. That day may indeed come. Still, the author is optimistic, viewing the final world as the recreation of a new heaven and a

Cosmic Evil and its Salvation

new earth. Those who fathom the purpose of the universe understand the meaning of this new creation well. However far humanity abandons itself to despair, it is unthinkable that the effort of cosmic will that has brought the universe this far in its evolutionary development will allow it to come to naught through the trifling failings of human beings. The grand path of creative evolution in the cosmos will move ahead quite well, I should think, even without humanity.

We are only halfway through cosmic evolution! The full production is yet to come! And here I sit watching it without even having paid the price of admission.

The dawn of the universe

Whether we search the inorganic world or inquire into the organic world, we find a design for the appearance of *life*, whose scale surpasses intelligence and law. One can only bow one's head at the ingenious planning of the universe.

The universe appears to be a careless collage of elementary particles, but when we look closely at the atoms that make it up, we cannot but find ourselves up against an *intelligence* hidden in the background. Even if the 238 atomic particles up to uranium are put together, the working capacity of the atoms is no more than twice that of hydrogen. Here again, we see the presence of strict regulations in the chaotic world of the atoms.

Mendeleev's periodic table of the atoms divided the atomic world into nine groups and the "Lewis-Langmuir principle" held that electrons concentrate a design for disposition and coordination in the eight corners of cubic cells. When we comprehend this, we are awestruck by the wonder of crystallization and the delicate workings of the laws of the universe.

We are in for a still greater surprise when we turn to Avogadro's world of molecular capacity. Even when all the atoms with all their differences combine, under the same temperature and atmospheric pressure, the space occupied by molecules is identical. This is a total miracle.

In order for life to appear, the universe has given birth to the magic of the proteins with their marvelous batteries. Set in a field with a positive electric charge, they effect a negative charge; and vice versa. Through the surprising make-up of organic fats, starches, enzymes, hormones, and vitamins, this magic has introduced living bodies to the earth.

These living bodies have been planned from the beginning for evolutionary development through the purposive management of genes and organizers. The process is controlled by laws of mitogenesis common to all living organisms, vertebrate and invertebrate alike. The skulls of vertebrates, from fish to mammals, show almost no change. Basically there are twenty-eight bones, and if we add the four extra bones in the ear, the skulls of mammals have thirty-two bones. The course that the evolution of the brain has taken is not by chance either; it is an orderly developmental path.

Embryological research on the generation of eggs shows an orthogenetic path of evolution, but in the world of nature in general the process is heterogenetic. The evolution of the "gait" of plants and animals represent an extreme form of heterogenesis, but when we study the dependency that exists between the two, we find that their respective struggles for survival do not lead to mutual extinction among life forms. On the contrary, we notice a strange, hidden versatility in the measures devised for living organisms to evolve.

Among Western biologists, the struggle for survival seems too cruel. Despite the many organisms that are born, the greater part of them are eaten by other animals. There is too much expenditure. Evolution is slow and riddled with imperfections. In particular, when we look at the female tarantulas and mantises that eat their male mates, it is hard to think there is purpose in nature. I have no wish to deny this, but would add that insects and spiders store the living flesh of the males that die after mating as food for their eggs or their offspring. Nevertheless, we have to recognize some points of difference between biologists looking for a simple way to tidy things up and the processes of nature.

From a human point of view, there is a great amount of bungling and imperfection to be found in the natural world. Taken as a whole, however, we may consider the imperfect parts more coherent than expected. Hidari Jingorō, the famous Tokugawa architect and sculptor, is said to have inserted the rafters in the Yōmeimon gate at Nikkō by deliberately turning them upside down. The appendix is said to serve no purpose for humans, and yet for ruminants it provides a useful place to breed microorganisms and digest plant fibers.

It will not do to judge survival and the struggle for survival in nature only from a human perspective. We cannot deny the presence of pain in the world, but neither can we look *only* at the pain or allow ourselves to deny the beautiful aspects and the movement of cosmic will to correct cosmic evil.

The drama that the universe is out to perform has not yet come to an end. Since the curtain has gone up, we are probably no further than the third act. Historically speaking, it has only been some twenty centuries since hope for the regeneration of human existence in the universe has come to our attention. There is much that lies ahead before the final curtain, and we must withhold judgment until the great finale.

But just because the raising of the curtain on an initial purpose is a thing of beauty and the intervening acts are also full of hope is no cause for a snap judgment on the cosmic conquest of evil. There is more to cosmic will than that.

I am thinking we have to await in hope the dawn of a universe that has earned the honor of overcoming the atomic bomb.

Cosmic evil and its salvation

The origins of evil are unknown and unidentified. From the viewpoint of cosmic purpose, the problem of the origins of evil is clear: it is the result of a failure to achieve cosmic purpose. Because cosmic purpose entails choices, even a slight breakdown in the conditions governing selection generates evil.

The occurrence of these slight breakdowns are unavoidable in a finite world. But there is heroism to be seen in the initial construction of our finite world, in the birth of *life*, in the emergence of *spirit* in the interior of life, and in the desire of the spirit to approach the unlimited and absolute.

Moreover, there is no forgetting the fact that in the intellectual history of humanity, solutions to liberation from evil have taken shape in religious beliefs.

In his *Divine Comedy*, the medieval Italian genius Dante resigned himself to our inability to answer the question of the final purpose of the cosmos:

> … for that which you ask lies so deep within the abyss of the eternal statute that it is cut off from every created vision. And when you return to the mortal world, carry this back, so that it may no longer presume to move its feet toward so great a goal. The mind, which shines here, on earth is smoky, and therefore think how it can do below that which it cannot do, though heaven raise it to itself. (*Paradiso*, canto 21)[2]

---

2. Dante Alighieri, *The Divine Comedy: Paradiso*, trans. by C. S. Singleton (Princeton: Princeton University Press, 1975), 239.

Such was Dante's surrender to the impenetrability of cosmic purpose, and I find I must agree with him.

However, as mentioned earlier, if we distinguish initial purpose from final purpose, we may yet be able to know something of its structure. On the basis of that judgment, we may make the following six statements:

1. There is purpose in the cosmos.
2. Cosmic purpose is directed towards *life*.
3. The purpose of life is directed towards *mind* (consciousness).
4. The individual mind is social and directed towards construction.
5. The social *mind* oriented to construction is en route to a historical evolutionary development and an awakening to cosmic consciousness.
6. This orientation awaits the assistance of the spirit that made creative evolution possible.

Along with the partial awakening of human consciousness, awareness of cosmic evil is deepened and the sense of the contradictions of human life deepens proportionally. We may liken it to passengers falling asleep on an express train. While asleep, they take no notice of changes around them, of the passage of time, of the crowded conditions. So, too, during those times when human beings are in the unconscious or subconscious state of mere mammals, they do not give a thought to cosmic evil and are incapable of awareness of their personal faults. Once they open their eyes, everything changes.

As passengers, the fact that they cannot go anywhere other than the determined destination becomes primary in their consciousness. They are not even permitted to ask about famous sites as the various stations along the route are being announced. They lack flexibility; change is restricted, as is time. They may only do the work they have chosen for that day. And then there are the regulations that cover their rail tickets. However free they may feel in their minds, as mammals they cannot surpass their animal sphere.

Put in these terms, the greater the spiritual ambitions of imperfect human beings, the deeper their consciousness of cosmic evil. To ignore our place in the middle of an ongoing cosmic evolution and imagine that we began as individuals from a state of untarnished perfection may be to trivialize the fact of being born human to the utmost degree.

A missionary to southern India writes that even one year spent living there in the world's most painful tropical regions makes one feel the impermanence of life as the Buddha had. But the sensation of pain is sharpened when adaptation to the natural environment and adaptation to the body's inner environment are mutually incompatible. Even when those who suffer are well medicated, they are still in pain. For those unable to see the whole picture and impatient for evolutionary development to come to an end, nothing is as insignificant as human life.

From ancient times people have set out to explain salvation from cosmic evil in one of three ways. First is India's religious way, the idea of emptiness. Second is the theistic approach to salvation that developed in western European thought. Third is the modern scientific attempt to banish cosmic evil.

I do not find these three to be incompatible. Each of them was bred in human consciousness. Nishida Kitarō recognized the conscious efficacy of the idea of "nothingness." In the middle ages, Nicholas of Cusa acknowledged "zero" algebraically. The modern quantum mechanic physicist Hermann Weyl has followed the same line of thought. We are right to eliminate the idea of a meaningless void, but I am speaking of opting for "zero" as a way to think of removing cosmic evil. Moreover, the third path of science's banishment of evil, in its modern meaning, also requires our utmost efforts.

There are, however, limits to human strength that leave us no other solution than to recognize the dependence of everything on an absolute cosmic will that has prepared, a priori, the strength for human beings to survive and for evolution to develop.

Besides discovering cosmic purpose, I believe we have to entrust development from here on to an absolute will that has bestowed it with purpose. If that is in fact the case, it makes no sense not to do so. I hold that awakened human consciousness should seek out the support of that transcendent cosmic will in its very struggle to bring everything out into the open.

Index of Personal Names

Aleksandrov, Georgiĭ Fedorovich, 38, 56, 59
Alighieri, Dante, 25, 267–8
Alverdes, Friedrich, 241
Amano Shigeyasu 天野重安, 160
Amemiya Eiichi 雨宮栄一, 7
Apollonius of Perga, 257
Aquinas. Thomas, 256
Araki Toshima 荒木俊馬, 99
Ariga Teru 有賀 輝, 245
Aristotle, 22, 43, 66, 95, 165, 255, 259
Arrhenius, Svante, 22, 39–40, 132
Aston, F. W., 71
Augustine, St., 101
Avogadro, Amedeo, 6, 55, 58, 67, 69–71, 102–3, 265
Azumi Hiroshi 安積宏, 62

Baldwin, Ralph B., 96
Banta, A. M., 185
Barry, J. M., 153
Barth, Karl, 18, 24,
Bateson, William, 162
Bergson, Henri, 3–4, 15, 38, 101–2, 162
Bernard, Claude, 157–8
Berzelius, Jöns Jacob, 67, 69
Bloom, Irene, 23
Bode, Johann Elert, 80–1, 231
Bohr, Niels, 33, 57, 72–4, 76-7, 93, 102
Bois-Reymond, Emil du, 215
Bonaparte, Napoleon, 262
Bondi, Hermann, 100
Bowne, Borden Parker, 15
Brachet, Jean, 151

Bradshaw, Emerson O., 4, 19
Bragg, William Henry, 81–2
Brahe, Tycho, 22, 39
Braun, Alexander Karl Heinrich, 86–9
Broglie, Louis de, 10, 84, 91, 93, 99–101
Brown, R. Hanbury, 98
Büchner, Ludwig, 100, 149
Bunge, Gustav Von, 113
Burbank, Luther, 43
Burns, David, 109–10
Butenandt, Adolf F. J., 59, 144, 167–8
Butterworth, Thornton, 39

Calvin, John, 23–4
Calvin, Melvin, 170–1, 243, 256
Cannon, Walter Bradford, 55, 160
Carrel, Alexis, 196, 232
Carus, J. V., 120
Compton, Arthur H., 10, 29, 211
Comte, August, 4
Copernicus, 96
Cori, Carl Ferdinand, 154
Coulter, John Merle, 174–5
Crick, F. H. C., 166, 168–9, 171–2
Crile, George Washington, 115, 140–3

Dalton, John, 67–9, 71
Dante. See Alighieri, Dante
Darwin, Charles, 3, 8, 10–13, 19, 21, 25, 53, 100–2, 104, 107, 109, 114, 149, 157–9, 162, 177, 186, 189, 203, 205–6, 209, 213–15, 223–4, 234, 252–5, 259
de Bary, Wm. Theodore, 23
De Sitter, Willem, 99, 101

Index of Personal Names

Democritus, 134
Dirac, Paul, 209
Disney, Walt, 186
Doppler, Christian, 100
Driesch, Hans 162–4, 237
Drummond, Henry, 241
Dürken, Bernhard, 162–6

Economo, Constantin Freiherr von, 244
Eddington, Arthur, 10, 101, 211
Edison, Thomas A., 261–2
Einstein, Albert, 3, 10, 79, 92, 99, 220
Elton, Charles, 181
Engels, Friedrich, 10, 102
Ernst, Frank A., 142
Eucken, Arnold, 4
Euclid, 91
Exman, Eugene, 16–17

Fabre, Jean-Henri Casimir, 179, 241, 260
Fairfield, Henry, 185
Faraday, Michael, 45, 58, 67, 71–2
Feather, Norman, 95
Fermi, Enrico, 73, 95
Fibonacci (Leonardo Pisano Bigollo), 86, 88
Fischer, H. Emil, 57
FitzGerald, G. F., 220
Fraenkel-Conrat, Heinz, 172
Freud, Sigmund, 251
Fröbel, Friedrich, 235
Fujita Tetsuo 藤田哲夫, 86
Fukurai Tomokichi 福来友吉, 251

Galilei, Galileo, 4
Gamow, George, 19, 85, 100–1, 103, 222
Gay-Lussac, Joseph Louis, 67–9, 220
George III, King, 97
Gibbs, J. Willard, 36, 47, 52
Gitlin, David, 146–7
Goudsmit, Samuel, 62, 219
Gray, George W., 152
Green, Eldridge, 245

Haas, Arthur, 74, 84
Harkins, William D., 66

Harrison, Ross G., 150
Hartman, Max, 176
Hashida Kunihiko, 219
Hazard, C., 98
Hecht, Selig, 245
Hegel, G. W. F., , 4, 63
Heidegger, Martin, 224
Heinrich, Emil, 215
Heisenberg, Werner, 10, 91, 211–12, 223, 250
Helmholtz, Hermann, 21, 100, 149, 213–14, 260
Henderson, Lawrence J., 41, 104–5, 107–9, 114, 121, 159
Herrick, Charles Judson, 249
Herschel, William, 97
Hidari Jingorō 左 甚五郎, 266
Hirth, Léon, 133
Holt, Henry, 249
Holtfreter, Johannes, 151
Honda Chikao 本田親男, 29
Hoyle, Fred, 19, 56, 101
Hubble, Edwin P., 80, 97–101, 220
Hull, Clark L., 251
Huxley, Julian, 13, 39, 79, 104, 108, 114, 159

Irving, Edward A., 52
Ishikawa Chiyomatsu 石川千代松, 34

James, William, 249, 251
Janeway, Charles A., 147
Jeans, James, 10–11, 80–1, 99, 101
Jeffreys, Harold, 96
Jesus Christ, 1, 6, 16, 20, 24, 242, 261
Jordan, Pascual, 93

Kagawa Haru 賀川ハル, 1, 6
Kagawa Jun'ichi 賀川純一, 7
Kandatsu Makoto 神立 誠, 178
Kant, Immanuel, 11, 23, 35, 43, 96, 102, 198–200, 255–6, 259
Kaō Kame 菅生かめ, 7
Kapteyn, Jacobus, 97
Katō Fusajirō 加藤普佐次郎, 251
Keller, Helen, 261

Index of Personal Names

Kepler, Johannes, 100, 257
Kesküla, Martin, 98
Kesküla, Tatjana, 98
Kihara Hitoshi 木原 均, 149
Kikuchi Masashi 菊池正士, 71
Kishi Hideshi 岸英司, 1, 3–4, 8, , 17–18, 25
Kobori Akira 小堀 憲, 257
Koga Shōzō 古賀正三, 159
Kornberg, Arthur, 171
Kozai Yoshishige 古在由重, 38
Kozinski, Andrzej, 162
Kraus, J. D., 98
Kropotkin, Peter, 4, 181
Kuroda Shirō 黒田四郎, 29

Lambert, Jean, 96
Lange, Friedrich Albert, 10
Langmuir, Irving, 75–6, 265
Laplace, Pierre-Simon, 198
Lawrence, Ernest O., 29, 100
Lecomte du Noüy, Pierre, 18, 42, 103
Leinen, Josef, 59
Lemaître, Georges, 99, 101
Lenard, Philipp, 116
Levinthal, Cyrus, 171
Lewis, Gilbert N., 265
Lindblad, Bertil, 97
Loder, James E., 7
Loeb, Jacques, 241–2
Lombroso, Cesare, 4
Lorentz, H. A., 220
Lotze, Hermann, 131
Lundmark, Knut, 98
Lysenko, Trofim Denisovich, 165

Malthus, Thomas, 100
Martin, D. H., 244
Marx, Karl, 4, 10–11, 102
Maxwell, James Clerk, 100, 230
Mayer, Maria Goeppert, 79, 95, 103
McCollum, Elmer Verner, 112–13, 153–4
McDougall, William, 250
McTaggart, H. A., 135
Mehl, Margaret, 18

Mendel, Gregor, 19, 83, 162, 190–1, 201, 207, 213
Mendeleev, Dmitri, 21, 46, 51, 68, 73, 75, 102, 265
Mihara Yōko 三原容子, 6
Millikan, Robert A., 10, 29
Minkowski, Hermann, 91
Mizuno Toshinojō 水野敏之丞, 29
Morgan, Thomas Hunt, 162, 201
Mori Kōchi 森 宏一, 38
Moseley, Henri, 67, 72, 83, 230
Mulder, Gerardus Johannes, 140
Murray, John, 252
Mutō Tomio 武藤富男, 17

Nagai Hisomu 永井 潜, 112–13
Nagel, Thomas, 20
Nakae Tōju 中江藤樹, 23
Nakamura Hiroshi 中村浩, 29, 154
Nakamura Seiji 中村清二, 85
Nakamura Shishio 中村獅雄, 10
Nakane Genkei 中根 元圭, 86
Nakaya Ukichirō 中谷宇吉郎, 116
Nakazawa Tsuneyuki 中澤恒幸, 146, 252
Napoleon. *See* Bonaparte
Naruse Masao 成瀬政男, 85
Needham, Joseph, 151–2
Newton, Isaac, 36, 99
Nicholas of Cusa, 5, 269
Niebuhr, Reinhold, 12
Nietzsche, Friedrich, 261
Nishida Kitarō 西田幾多郎, 5, 269
Nissl, Franz, 115, 141–2
Norman L. Bowen, 52
Numata Makoto, 152

Ōyōmei. *See* Wang Yangming
Okamura Kintarō 岡村金太郎, 173
Ono Shun'ichi 小野俊一, 13
Oort, Jan, 97
Oparin, Aleksandr, 56–7, 59, 132, 134–6, 144–5, 149, 170, 201, 238, 256
Osborn, F., 185
Ostwald, Wolfgang, 104, 197, 230, 235

Index of Personal Names

Partington, John S., 13
Pasteur, Louis, 132
Paul, St., 260
Pauli, Wolfgang, 47, 208
Pauling, Linus, 29, 44, 62, 70, 102, 110, 209–10, 212, 219, 222
Perrin, Jean, 58
Pfeffer, Wilhelm, 95, 152
Pflüger, Eduard, 254
Pickering, E. C., 122
Planck, Max, 66–7, 81, 84–5, 212, 227, 230, 235, 250
Plato, 22, 66
Poulton, E. B., 214
Pradesh, Himachal, 116
Pröscholdt, Hilda, 151
Prout, William, 67, 71, 263
Ptolemy, 22, 40, 66
Purkinje, Jan Evangelista, 141–3

Redi, Francisco, 119
Reischauer, Edwin O., 17–18
Reynolds, J. H., 98
Rhine, J. B., 251
Rich, Alexander, 171
Ringer, Sidney, 113
Ritchey, George Willis, 97
Ritz, Walter, 62
Ruskin, John, 15
Russell, Bertrand, 4, 13, 39, 149–50
Russell, Henry Norris, 96, 232
Rutheford, Ernest, 74
Rydberg, Johannes, 62

Sachs, Julius, 95, 152
Saitō Kazuo 斎藤一雄, 152
Sakaguchi Kin'ichirō 坂口謹一郎,, 136
Sanger, Frederick, 140
Scharrer, Ernst, 243–4, 252
Schildgen, Robert, 16
Schimper, Karl Friedrich, 86–9
Schleiermacher, Friedrich, 4
Schopenhauer, Arthur, 43, 238
Schotte, Oscar E., 152
Schramm, Gerhard, 172
Schrödinger, Erwin, 10, 91, 157

Schwann, Theodor, 252
Shakespeare, William, 38, 65, 145
Shakyamuni, 189
Shapley, Harlow, 97, 99
Shaw, George Bernard, 38, 162
Sherman, Mildred S., 142
Shōji Rinnosuke 正路倫之助, 113
Shushi. *See* Chu Shi
Siekevitz, Philip, 146
Simpson, George, 116
Singleton, C. S., 267
Sitter, Willem de, 101
Smuts, Jan, 163
Sommerfeld, Arnold, 43, 57–8, 76, 102, 190, 209–10, 212, 222, 230, 234–6, 256
Spemann, Hans, 150–1, 237, 252
Stent, Gunther S., 172
Stokes, George Gabriel, 76
Stolkowski, Joseph, 133
Stoney, George J., 58
Strömberg, Gustaf, 18, 237
Suess, Hans, 118
Sugiura Yoshikatsu 杉浦義勝, 209
Sumiya Mikio 隅谷三喜男, 17–18
Sumner, James, 135
Suzuki Keishin 鈴木敬信, 101
Swedenborg, Emanuel, 96, 261
Szent-Györgyi, Albert, 63–4, 139, 145, 147–9

Takahashi Jun'ichi 高橋純一, 117
Takahashi Masami 高橋正躬, 178
Taketani Mitsuo 武谷三男, 92
Takeuchi Masaru 武内 勝, 198
Tamamushi Bun'ichi 玉蟲文一, 135
Tanaka Minoru 田中 実, 58, 71
Taylor, J. Herbert, 172
Teilhard de Chardin, Pierre, 3–4
Teubner, B. G., 163
Thales, 109, 257
Thompson, E. O. P., 140
Tiselius, Arne, 147
Tolman, Edward Chace, 251
Tolstoy, Leo, 15
Troeltsch, Ernst, 259

274

Index of Personal Names

Vaucouleurs, Gérard de, 98-9
Vernadsky, Vladimir, 58, 104, 117-22, 179
Vries, Hugo de, 162, 204

Waddington, C. H., 151
Wallace, Alfred Russel, 19, 162, 241
Wang Yangming 王陽明, 23
Warburg, Otto, 142-3
Watson, James D., 168
Watson, John B., 249-51
Wells, G. P., 13
Wells, H. G., 12-17, 26, 38-9, 50, 65, 79, 112, 134, 145, 150, 158, 165-6, 204, 256
Went, Fritz W., 183

Weyl, Hermann, 5, 84, 92, 269
Wheeler, William, 241, 263
Woodworth, Robert S., 249
Woolley, Dilworth Wayne, 161
Wright, Thomas, 96
Wundt, Wilhelm, 249

Yamao Yasumasa 山尾泰正, 219
Yamashita Ryūji 山下竜二, 23
Yoshimoto Kōzō 吉本浩三, 29
Yoshimura Hisato 吉村寿人, 113
Yukawa Hideki 湯川秀樹, 72, 85, 90-3

Zhu Xi 朱熹, 23
Zhu Zaiyu 朱載堉, 86

www.ingramcontent.com/pod-product-compliance
Lightning Source LLC
Chambersburg PA
CBHW030612230426
43661CB00053B/1958